Assessing the Impact of Future Operations on Trainer Aircraft Requirements

John A. Ausink, Richard S. Marken, Laura Miller,
Thomas Manacapilli, William W. Taylor,
Michael R. Thirtle

Prepared for the United States Air Force

PROJECT AIR FORCE

The research described in this report was sponsored by the United States Air Force under Contract F49642-01-C-0003. Further information may be obtained from the Strategic Planning Division, Directorate of Plans, Hq USAF.

Library of Congress Cataloging-in-Publication Data

Assessing the impact of future operations on trainer aircraft requirements / John A. Ausink ... [et al.].
p. cm.
"MG-348."
Includes bibliographical references.
ISBN 0-8330-3790-0 (pbk.)
1. United States. Air Force—Procurement—Evaluation. 2. Training planes. I. Ausink, John A.

UG1123.A72 2005
358.4'183—dc22

2005007838

The RAND Corporation is a nonprofit research organization providing objective analysis and effective solutions that address the challenges facing the public and private sectors around the world. RAND's publications do not necessarily reflect the opinions of its research clients and sponsors.

Published 2005 by the RAND Corporation
1776 Main Street, P.O. Box 2138, Santa Monica, CA 90407-2138
1200 South Hayes Street, Arlington, VA 22202-5050
201 North Craig Street, Suite 202, Pittsburgh, PA 15213-1516
RAND URL: http://www.rand.org/
To order RAND documents or to obtain additional information, contact
Distribution Services: Telephone: (310) 451-7002;
Fax: (310) 451-6915; Email: order@rand.org

Preface

This report documents the RAND Corporation's research on how the skills required in future U.S. Air Force aircraft might affect the decision to retain or replace current trainer aircraft. It is based on over 200 interviews with Air Force pilots in a wide variety of flying assignments, a review of Air Force documents on future flying missions, and discussions with flying training experts in other countries. This research, under the title "Flying Training 2020," was sponsored by General Donald G. Cook, the commander of Air Education and Training Command (AETC), and performed within the Manpower, Personnel, and Training Program of RAND Project AIR FORCE.

This report is designed to help AETC and the Air Force make informed decisions about retaining or replacing current trainer aircraft in order to best prepare pilots for the aircraft they will fly through the year 2040.

RAND Project AIR FORCE

RAND Project AIR FORCE (PAF), a division of the RAND Corporation, is the U.S. Air Force's federally funded research and development center for studies and analyses. PAF provides the Air Force with independent analyses of policy alternatives affecting the development, employment, combat readiness, and support of current and future aerospace forces. Research is conducted in four programs: Aerospace

Force Development; Manpower, Personnel, and Training; Resource Management; and Strategy and Doctrine.

Additional information about PAF is available on our Web site at http://www.rand.org/paf.

Contents

Figures

Tables

Summary

Developing the requirements and securing the funding for modern military aircraft can take a significant amount of time. Given emerging operational demands and the age of some current Air Force trainer aircraft, it is time to examine how the skills needed to perform future military missions might affect the capabilities required of new aircraft and ground-based systems used in pilot training.

From 1962 until 1992, Air Force pilots learned to fly in an Undergraduate Pilot Training (UPT) program in which all students first flew the subsonic T-37 jet aircraft and then the supersonic T-38. In 1992 the Air Force began a transition to Specialized Undergraduate Pilot Training (SUPT), which tracked students after the T-37 phase of training. Students selected to fly fighters or bombers now train in the T-38, while those selected to fly tanker or transport aircraft train in the T-1A, a military derivative of a commercial business jet. The Air Force began replacing the T-37 with a new aircraft in 2001, but in the next few years, it must make decisions to replace or extend the lives of the aging T-38 and the newer, but tiring, T-1A.

The timing of these decisions is important because the inventory of Air Force aircraft will change dramatically in the next 25 years. Two new fighter aircraft, the F/A-22 and the F-35, will be introduced, with the F/A-22 replacing the F-15 and F-117 and the F-35 replacing F-16s and A-10s. While there are no plans to develop new transport, tanker, or bomber aircraft over the next two decades,[1] pilots of these aircraft, like fighter pilots, will face a future that is characterized by the following:

- operations conducted around the clock and in all weather and geographical conditions (pp. 30–31)
- operations requiring near real-time implementation of airpower against an enemy (pp. 31–32)
- incorporation of precision weapons to increase mission effectiveness while minimizing the exposure of manned aircraft to threats (pp. 32–33)
- mobility missions taking place in closer proximity to the enemy (pp. 34–35)
- integration of large amounts of information from disparate sources (land, air, and space based) in real-time conditions (pp. 35–36)
- flight profiles involving greater physiological demands (pp. 41–43).

More complicated missions, new aircraft capabilities, and new information management demands require new pilot skills. The question is, which of these skills, if any, should be taught in undergraduate flying training, and which, if any, are so different that they cannot be taught in current training aircraft? If required future skills are beyond the capabilities of current training aircraft, the decision to replace them is obvious. If not, a service life extension program (SLEP)[2] of the aircraft might be acceptable.

The RAND Corporation was asked by Air Education and Training Command (AETC) to examine the replacement decision for

---

[1] The Air Force has been examining the possibility of replacing its tanker fleet, but the potential replacements are derivatives of existing commercial aircraft. The C-130J can be considered a new transport aircraft, but it is a modification of an airframe already in the inventory.

[2] A SLEP is a modification to an aircraft that is made to extend the life of the aircraft beyond what was originally planned.

the T-38C and the T-1A, and to do so we reviewed Air Force planning documents related to future approaches to combat, attended an international conference on pilot training, and discussed future aircraft inventories with officers in the planning business. More important, we interviewed 230 Air Force pilots involved in every stage of the pilot training pipeline and representing experience in virtually all current Air Force aircraft to hear their opinions about current pilot training and their predictions about the skills pilots will need in the future. Despite the diverse backgrounds of those we interviewed, we saw a convergence of themes that indicated to us that pilots perceived the same types of challenges and recommendations for flying training across the organization. Among these are the following:

- The collection, synthesis, and prioritization of information in the cockpit will become increasingly difficult in future operations as we look out toward 2025 (pp. 37–38).
- Flying the aircraft is currently and must continue to be second nature for pilots, given the many information management tasks that are prevalent in the operational environment (pp. 43–48).
- Pilots will continue to be challenged with more responsibilities in the cockpit that are focused on the management of information, sensors, and weapons (pp. 43–48).
- Even with the changes in technology and the impact of such changes in the operational flying environment, SUPT should continue to focus on teaching flying fundamentals to new pilots (pp. 43–48).
- Pilots will be required to become more proficient at layering technological solutions in the cockpit, that is, knowing when, and when not, to make use of (or depend on) a given technology (pp. 43–48).

Recognizing these training themes, the almost unanimous conclusion of the pilots we interviewed was that the aircraft currently used in SUPT are capable of providing the skills required of pilots who will be flying Air Force missions over the next several decades.[3] Thus, our first conclusion in this study is that in the context of SUPT, the decision to replace, or extend the life of, the T-38 and the T-1A can be reduced to an economic analysis based on cost alone: do whichever is cheaper.

Our discussions with pilots showed, however, that the replacement decision must be based on an analysis of the entire training pipeline, and not just SUPT. For example, SUPT graduates who are assigned to the F-15 and F-16, both single-seat fighters, first gain some experience in two-seat versions of the aircraft at their Formal Training Units (FTUs),[4] but there are no plans to develop two-seat versions of the F/A-22 or the F-35. G-induced loss of consciousness (GLOC) is not unusual during training in the F-15 and F-16, but new students are protected by the presence of an instructor.[5] Pilots we interviewed were concerned about the potential dangers of exposing inexperienced SUPT graduates to the high-g capabilities of the F/A-22 and the F-35 when their first flight in those aircraft will be solo. Some suggested the use of an intermediate training aircraft to ease the transition from the Introduction to Fighter Fundamentals (IFF) course (taken by SUPT graduates assigned to fighter aircraft) to the F/A-22 or the F-35. Implementing that suggestion could have a ripple effect on what is taught in SUPT, the demands on aircraft, and the decision to replace the T-38.

Several categories of issues could increase or decrease the demands on current training airframes or affect the type of training pi-

---

[3] Current SUPT aircraft are the T-6, which is replacing the T-37, the T-1A, the T-38A, and the T-38C. The T-38C is currently flown in SUPT at Vance AFB, Oklahoma, and Columbus AFB, Mississippi, and in Introduction to Fighter Fundamentals (IFF) training at Moody AFB, Georgia. Eventually all T-38 aircraft used in SUPT will be upgraded to the T-38C.

[4] FTUs are where pilots first learn to fly their assigned aircraft.

[5] G-forces are the forces of acceleration experienced by a pilot while maneuvering in flight. A force of nine positive g's makes a pilot feel nine times as heavy as when the aircraft is in unaccelerated flight. High positive g-forces can reduce the flow of blood to the brain, which can lead to loss of consciousness.

lots need, and thus influence decisions on the feasibility of retaining current aircraft:

- **Strategy**: The demand for pilots could change as a result of the increased use of Unmanned Aerial Vehicles (UAVs) or other decisions that affect the number of aircraft needed to perform Air Force missions. If fewer pilots are needed, fewer will need to be trained, and the demands on current trainer aircraft will be reduced. UAV operators may need new approaches to training.
- **Policy**: A desire for increased flexibility in the assignment of Air Force pilots could lead to the requirement that all pilots receive the same training. In addition, moves to increase joint operations with other services may affect the amount of training required.
- **Training:** More advanced simulators and other improved approaches to ground-based training could affect the number of flying training hours required. Any changes in the timing of the tracking decision will also affect the demands on the T-38. Changes in the requirements for FTUs could also affect the decision to retain or replace current training aircraft.
- **Budget:** Better understanding of the economics of aging aircraft will affect the decision, since in the case of the T-38, retaining the aircraft will mean that the Air Force could eventually be training student pilots in jets close to 80 years old. In addition, any decision about a follow-on trainer aircraft must take into account the costs associated with classroom instruction, computer-based training, simulators, and other ground-based training necessary to augment what is taught in the aircraft.

Therefore, while the T-1A and the T-38 are adequate for future training in the context of SUPT, AETC should consider using the following approach to finalizing its replacement decision:

1. As a baseline, determine the cost of continuing SUPT and IFF in their current forms by SLEPing the T-38C and the T-1A. At the same time, determine the cost of retaining trainer versions of the F-16 in order to use them in a pre–F/A-22 FTU program that will expose new pilots to high sustained g-forces in the presence of an instructor before they fly solo in the F/A-22 (pp. 51–52, 54).
2. Compare this to the cost of continuing SUPT in its current form but with replacement aircraft for the T-1A and the T-38C. For T-38C replacement aircraft, it makes sense to consider some version of the BAE Hawk (already used in training by other air forces) or the T-50 (recently developed for training in the Korean Air Force). This comparison should also include the cost of using the replacement for the T-38C in IFF and in a pre–F/A-22 FTU program (pp. 61–62).
3. As a first excursion from current SUPT, consider the possibility of extending T-6 training before the tracking decision is made (pp. 21–25). This would decrease the demands on both the T-38C and the T-1A in SUPT, which could mean they would last longer even without a SLEP.
4. As a second excursion from current SUPT, the effect of allowing all students to fly the T-38C (in order to expose them to a higher performance aircraft before tracking) should be analyzed (pp. 21–25).[6]
5. Finally, examine the costs of returning to single-track UPT, first in a version using a SLEPed T-38C,[7] and then with a replacement

---

[6] Lt Gen Baker, Vice Commander of Air Mobility Command (AMC), and others have suggested that all students have as many as eight sorties in the T-38C before tracking is done.

[7] Chocolaad (2001) considered this case in an excursion of his study, and determined that the Air Force would have to purchase T-38s from other U.S. (and perhaps foreign) organizations in order to meet the expanded sortie requirements. Because of attrition, he concluded that the T-38 would not be able to meet sortie requirements after 2020. The study took into

for the T-38C. In the second case, the replacement aircraft would be used for both the advanced training phase and IFF, and possibly for a pre–F/A-22 version of FTU. The single-track option would also introduce a tanker/transport version of IFF: after single-track UPT, graduates could attend a short course in the T-1A before going on to FTU (many people we interviewed think that current T-1 training is too long). This option will be interesting to consider, because it would replace only one aircraft (instead of replacing or SLEPing two), and might provide an option for interim training for new pilots who have been assigned to the F/A-22. Thus, while there is no compelling training reason to return to single-track UPT, cost considerations or increased flexibility in making pilot assignments might make doing so an attractive option (pp. 61–62).

account the costs of some upgrades to the T-38 (including those for the T-38C), but it did not consider a more extensive SLEP that might make the aircraft more reliable or sustainable.

Acknowledgments

This project would not have been possible without the extraordinary cooperation of many Air Force organizations. We would like to thank, first of all, Lt Col Bud Brooks and Mr. Rick French, our project officers in the Undergraduate Flying Training Requirements Support Branch (AETC/XPRU), who made a tremendous effort to establish points of contact at more than a dozen training, operational, and headquarters bases. These points of contact recruited pilots for us to interview, scheduled interviews, and provided logistical support during our visits; their willingness to help despite the administrative and flying demands of their real jobs made it possible for us to complete our research in the short period of time required by our sponsor, and we are grateful for their assistance. By command, these officers were the following:

HQ USAF: Maj Tom Daack (XPX) and Lt Col Rand Miller (XPXC)

ACC: Maj John Farese (Barksdale AFB), Lt Col Mike Senna (Holloman AFB), Lt Col Mark Hamilton (Langley AFB), Capt Jason Trew (Langley AFB), Capt Chuck Michalec (Nellis AFB), and Capt Matt Brechwald (Whiteman AFB)

AETC: Lt Col Bryan Riba (Altus AFB), Capt Andy Shields (Kirtland AFB), Capt Chad Del Rossa (Laughlin

	AFB), Maj Ken Smith (Luke AFB), Lt Col Kirk Horton (Moody AFB), and Lt Col Pete Lasch (Randolph AFB)
AFMC:	1st Lt Richard Baysinger (ASC/YTG)
AFSOC:	Maj Mike Lewis and Lt Col Bob Monarch (Hurlburt Field)
AMC:	Lt Col Brian Wilson and Maj Bill Nietzke (Scott AFB)

From June 2004 to August 2004, 230 pilots were interviewed for this project. All of them provided candid assessments of the graduates of Specialized Undergraduate Pilot Training (SUPT) and shared thoughtful insights into the skills that Air Force pilots will need in the future. Their cooperation was the key to this research, and we regret that for reasons of confidentiality we are unable to thank them here by name. We were impressed by the professionalism and dedication of all the pilots we interviewed.

We benefited from the support of several general officers during our research. Our sponsors, Gen Donald Cook, Commander of AETC, and Brig Gen Gilmary Hostage, Director of Plans and Programs, encouraged the participation of non-AETC bases in this study. Lt Gen John Baker, Vice Commander of AMC, gave generously of his time during interviews at Scott AFB, and Lt Gen Bruce Wright, Vice Commander of Air Combat Command (ACC), allowed us access to several offices at Langley AFB.

We also thank Jack Graser, who assisted in conducting interviews; Al Robbert, Skip Williams, and Lt Col Scott Davis, who provided helpful comments on an early draft of this document; and Lisa Price, who edited the final manuscript.

Acronyms

AATT	Aging Aircraft Technologies Team
ACC	Air Combat Command
AETC	Air Education and Training Command
AETC/XPRU	Undergraduate Flying Training Requirements Support Branch
AFIT	Air Force Institute of Technology
AFCIS	Air Force Capability Investment Strategy
AFMC	Air Force Materiel Command
AFRL	Air Force Research Laboratory
AFSC	Air Force Systems Command
AFSOC	Air Force Special Operations Command
AFTTP	Air Force tactics, techniques, and procedures
AMC	Air Mobility Command
AMI	advanced manned interceptor
AMW	Air Mobility Wing
AOA	angle of attack
ASC	Aeronautical Systems Center
ATC	Air Training Command
AUP	Avionics Upgrade Program
BW	Bomber Wing
CAF	Combat Air Force

CAS	close air support
CONOPS	concept of operations
COTS	commercial off-the-shelf
CRM	cockpit/crew resource management
CSAF	Chief of Staff of the Air Force
DMO	distributed mission operations
DSL	design service life
ENJJPT	Euro-NATO Joint Jet Pilot Training
FAA	Federal Aviation Administration
FAIR	fighter-attack-intercept-reconnaissance
FEB	Flying Elimination Board
FL	flight level
FTG	Flying Training Group
FTU	Formal Training Unit
FTW	Flying Training Wing
FW	Fighter Wing
GAO	Government Accountability Office
GLOC	g-induced loss of consciousness
GPS	global positioning system
HOTAS	hands-on throttle and stick
HQ AF/DP	HQ U.S. Air Force, Deputy Chief of Staff, Personnel
HUD	head-up display
ICAO	International Civil Aviation Organization
IFF	Introduction to Fighter Fundamentals
IFT	Introductory Flying Training
IP	instructor pilot
ISR	intelligence, surveillance, and reconnaissance
JDAM	Joint Direct Attack Munition
JSF	Joint Strike Fighter

JSOW	Joint Stand-Off Weapon
JSUPT	Joint Specialized Undergraduate Pilot Training
KAI	Korean Aerospace Industries
MAJCOM	major command
MTC	Mission Training Center
MWS	major weapon system
NAS	Naval Air Station
NASA	National Aeronautics and Space Administration
NVG	night-vision goggle
PAF	Project AIR FORCE
PFT	programmed flying training
PGM	precision-guided munition
PME	professional military education
PMP	Propulsion Modernization Program
POM	Program Objective Memorandum
RAF	Royal Air Force
RPA	remotely piloted aircraft
RVSM	reduced vertical separation minimums
SA	situational awareness
SAC	Strategic Air Command
SAT	surface air attack
SDB	small-diameter bomb
SLEP	service life extension program
SMO	System Management Organization
SOF	Special Operations Forces
SOW	Special Operations Wing
SPO	System Program Office
SUPT	Specialized Undergraduate Pilot Training
TAC	Tactical Air Command

TCAS II	Traffic Alert and Collision Avoidance System II
TTB	tanker, transport, and bomber
UAV	unmanned aerial vehicle
UCAV	unmanned combat aerial vehicle
UHF	ultra-high frequency
UPT	Undergraduate Pilot Training
VHF	very high frequency
VTOL	vertical takeoff and landing

CHAPTER ONE

Introduction

Background: Air Force Pilot Training[1]

Several phases of training and a variety of aircraft and other types of training equipment are required to produce an Air Force pilot ready to perform a peacetime or wartime mission; Figure 1.1 is a schematic of this Air Force training pipeline. The undergraduate portion of training consists of three phases: screening, primary, and advanced training. During the screening phase, pilot training candidates complete 50 hours of Introductory Flying Training (IFT), which determines a candidate's suitability for selection into the primary phase of training. Civilian instructors conduct flight screening around the country. Some graduates of IFT who will fly fighter aircraft attend Euro-NATO Joint Jet Pilot Training (ENJJPT) at Sheppard AFB, Texas,[2] but most students go on to SUPT or Joint SUPT (JSUPT) programs, conducted at four Air Force bases and one naval base.[3]

[1] Much of the information in this section is from a U.S. Air Force fact sheet (2003b).

[2] ENJJPT trains students from eight countries: Belgium, Denmark, Germany, Italy, the Netherlands, Norway, Turkey, and the United States. At ENJJPT, all students fly the T-37 in primary training and the T-38 in advanced training. All graduating pilots are assigned to fighters, so there is no tracking after the T-37 phase of training. From FY 1999 to FY 2002, an average of 10 percent of new Air Force pilots graduated from ENJJPT (Air Education and Training Command, 2003c).

[3] SUPT is taught at Columbus AFB, Mississippi; Laughlin AFB, Texas; and Moody AFB, Georgia. JSUPT is taught at Vance AFB, Oklahoma, where some of the students are officers

Figure 1.1
Schematic of Joint Specialized Undergraduate Pilot Training

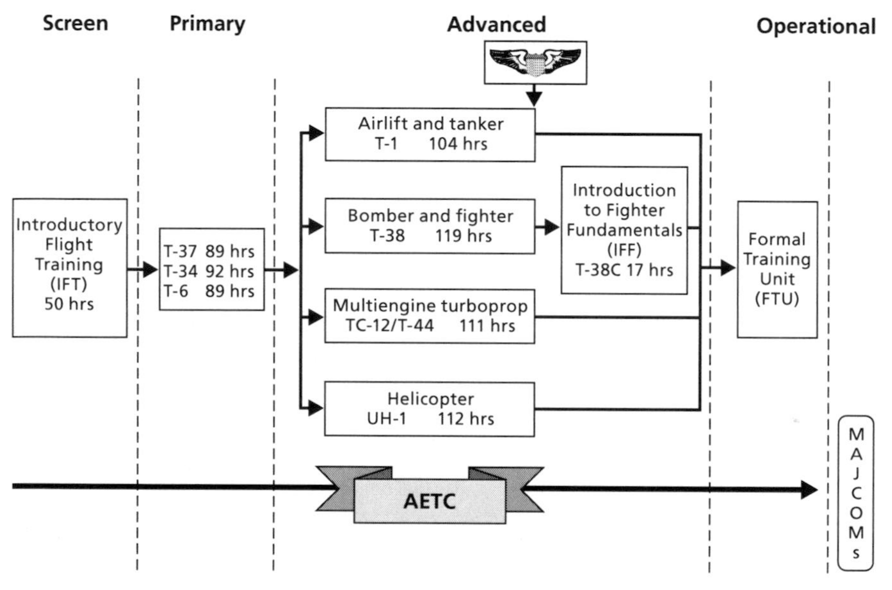

SOURCE: Fraser, 2004.
RAND *MG348-1.1*

The primary stage of SUPT is conducted in one of three aircraft: the T-37, a twin-engine subsonic jet aircraft with side-by-side seating; the T-6, a single engine turboprop with tandem seating; or, at Whiting Field, the T-34C, which is an older airframe than the T-6 but also a single engine turboprop. The Air Force has been using the T-37 in pilot training since 1956 (the T-37B version became the standard in 1959), and the Navy has flown the T-34C since the mid-1970s. The T-6, which was introduced in October 2001, will eventually replace both the T-37 and the T-34C so that all SUPT students will fly the same aircraft in the primary phase.[4] Primary training introduces student pilots to basic aircraft handling, instrument flying, two-ship formation, and basic navigation.

in the Navy. As part of the JSUPT program, some Air Force student pilots are trained with the Navy at Whiting Field, Florida.

[4] The Air Force will phase out the T-37 in about 2008.

After the primary phase of training, students are tracked—split into different training groups for different aircraft—for the advanced phase of SUPT based on their performance in training, instructor recommendations, a student's personal preference, and available aircraft assignments. Helicopter students go to Fort Rucker, Alabama, for advanced training with the Army, and students who will fly versions of the C-130 attend training with the Navy at Corpus Christi Naval Air Station (NAS) in Texas. Students selected for other airlift or tanker positions continue their advanced training in the T-1A, a military version of a commercial business jet, in which students learn how to manage a crew and are exposed to techniques for aerial refueling and airdrop missions. Finally, students who will be assigned to fighter or bomber aircraft have advanced training in the T-38,[5] a supersonic trainer with tandem seating, where they focus on two- and four-ship formation, low-level missions, and more instrument and navigation training.

Students are awarded their pilot wings after successfully completing the advanced phase of training, having spent approximately 52 weeks in SUPT. Pilots assigned to fighter aircraft then attend a short (about 40 days) Introduction to Fighter Fundamentals (IFF) course before moving on to Formal Training Units (FTU) where they begin training in their assigned aircraft.[6] Nonfighter pilots do not have any other flying training between graduation from SUPT and the beginning of training at their FTU.

As indicated by the arrow at the bottom of Figure 1.1, AETC manages training from IFF through some FTUs,[7] and then pilots are assigned to their major commands (MAJCOMs).

We will discuss the history of pilot training in more detail in Chapter Two, but we should note here that SUPT was initiated at Reese AFB in July 1992, and was the Air Force's approach to pilot

---

[5] This could be either the T-38A or the T-38C, depending on the training location.

[6] Fighter-bound students attend IFF at Moody AFB in the T-38C or at Sheppard AFB in the AT-38B, depending on availability and training quotas.

[7] The FTU for F-15Es is managed by ACC at Seymour Johnson AFB, and the FTU for A-10s is managed by ACC at Davis-Monthan AFB.

training at all pilot training bases by 1997 (U.S. Air Force, April 2003b). From February 1962 until 1992, all Air Force pilots had been trained in a generalized undergraduate pilot training program (UPT) in which all students flew the T-37 and the T-38.

The Potential Need for New Trainer Aircraft

As mentioned above, the T-37 aircraft flown in the primary phase of pilot training has been used by the Air Force for over 40 years, and is being gradually replaced by the new T-6. The T-38 began service in Air Force pilot training in 1961, and about 1,100 of the aircraft were purchased through 1972, when the production line was closed (U.S. Air Force, 2003c). The T-38 has been modified several times in its years of service, the most recent and extensive changes being those for the T-38C.[8] This version of the aircraft is now flown in the advanced phase of instruction for fighter and bomber pilots at two SUPT locations,[9] as well as in the IFF course at Moody AFB. Current projections by the Flight Training System Program Office (SPO) at Wright-Patterson AFB show the T-38C reaching the end of its service life in 2020. Projections for the much younger T-1A, which was introduced in SUPT in 1992, show that it will also reach the end of its service life in 2018 (Air Education and Training Command/ XPPX, 2004a, 2004b). Figure 1.2 displays the decisions facing the Air Force for these two aircraft.

[8] The Avionics Upgrade Program (AUP) gives the T-38C glass instrument displays, a head-up display (HUD), and integrated digital avionics. The Propulsion Modernization Program (PMP) makes modifications to the engines that increase thrust and improve takeoff performance, save fuel in some training profiles, and improve range slightly (AETC, undated).

[9] The SUPT locations are Vance AFB, Oklahoma, and Columbus AFB, Mississippi. The T-38C will soon also be flown at Laughlin AFB, Texas. ENJJPT will start using the T-38C in 2005.

Figure 1.2
Replacement Decisions for the T-38 and the T-1

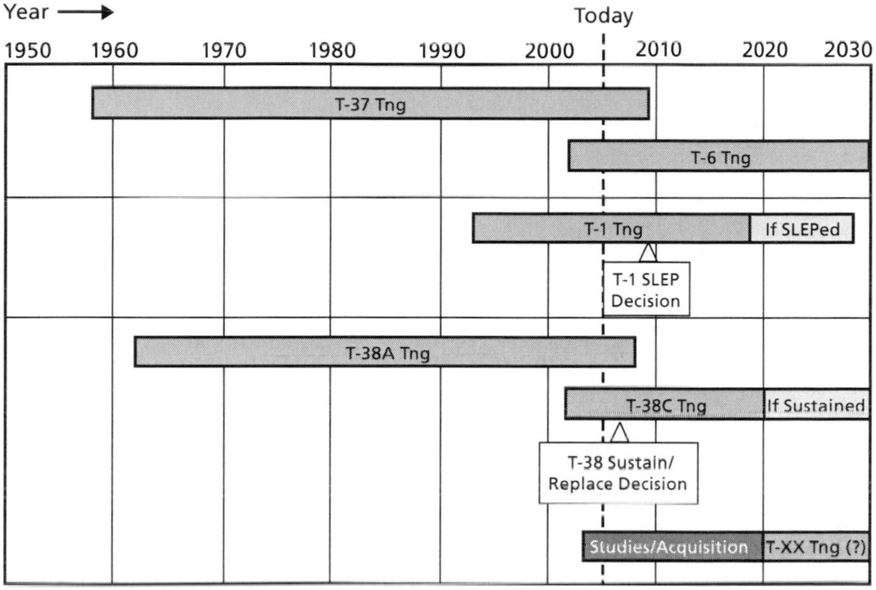

SOURCE: Palumbo, 2003.
RAND *MG348-1.2*

The chart indicates the service lives of the T-37, the T-6, the T-1A, the T-38A, and the T-38C. In 2004, we see the T-37 still in service, but the T-6 is also being flown at some bases. Use of the T-38 overlaps with the T-38C shortly after 2000, when the T-38C was introduced. The T-1A's projected service life ends in 2018, with the potential for use to 2028 if the aircraft is reconditioned as part of a SLEP. The chart also shows that the T-38C could be extended beyond 2030 (perhaps to 2040) if it is reconditioned by a SLEP as well. If these aircraft are not SLEPed, they will need to be replaced.

The figure shows the dates by which AETC Plans and Programs experts think the decision to SLEP or replace each aircraft must be made in order to ensure that training can continue beyond the year

2020: about 2006 for the T-38C, and 2009 for the T-1A.[10] The timing of these decisions, and the appropriate decisions to make, are affected by coming changes in the Air Force's aircraft inventory.

Air Force Aircraft Inventory

The average age of all Air Force aircraft in FY 2002 was 22 years, with 46 percent of all aircraft being older than 21 years (U.S. Air Force, 2003d, p. 83). The oldest operational aircraft are B-52s (average age 40.8 years in 2002) and KC-135s (average age 40.7 years), while fighter aircraft are younger: the average F-15 is 16.4 years old, and the average F-16 is 11.6. The Government Accountability Office (GAO) has noted that aging aircraft, especially tactical aircraft, have the potential to drive up operations and support costs (U.S. Government Accountability Office, 1989, pp. 2–3), and the Air Force has been planning for several years to introduce two new fighters: the F/A-22 and the F-35 (Joint Strike Fighter). The acquisition of these new aircraft and the decision to retain older mobility and tanker aircraft beyond their originally intended life spans will have a great impact on the composition of the aircraft fleet, and Figure 1.3 shows potential changes in the Air Force inventory over the next 20 years.[11] We must first note that some of the numbers predicted by the model used to create the FY 2025 values in the figure are overly optimistic. There is great uncertainty over the number of F/A-22s that will ultimately be purchased, for example, and recently published estimates point to a procurement potentially as low as 217 (Ahearn, 2004). Also, while original plans called for a one-for-one replacement of F-16s with Joint Strike Fighters, the actual number purchased will likely be less, and the Air Force will not specify how many it plans to buy until the FY 2006 Program Objective Memorandum (POM) ("Jumper," 2004).

---

[10] The T-38C decision must be made earlier because of the possible requirement to develop a new aircraft. A T-1A replacement will likely be another off-the-shelf aircraft.

[11] The Air Force has been examining the possibility of replacing its tanker fleet, but the potential replacements are derivatives of existing commercial aircraft. The C-130J can be considered a new transport aircraft, but it is a modification of an airframe already in inventory.

Figure 1.3
Potential Changes in Air Force Aircraft Inventory from FY 2002 to FY 2025*

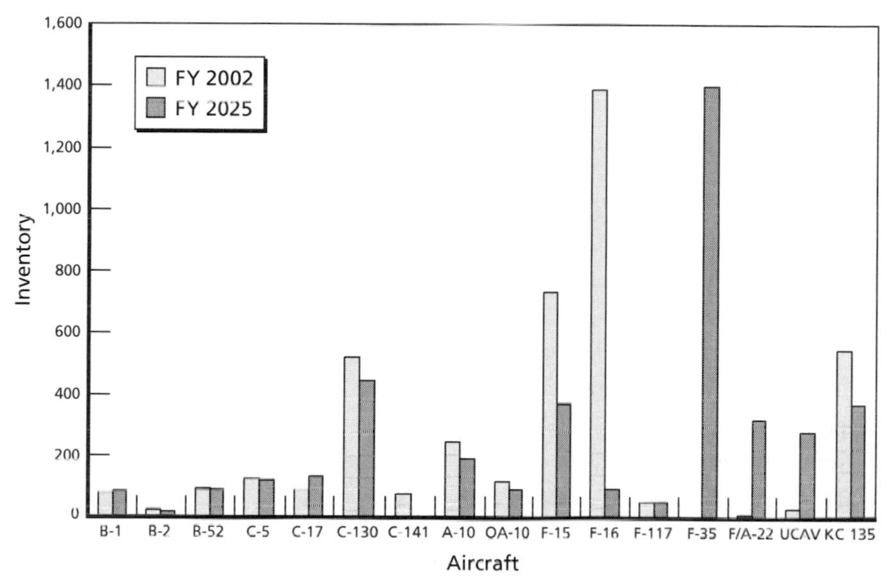

*NOTE: FY 2002 inventory numbers are from the 2003 Air Force Almanac (U.S. Air Force, 2003d). Figures for FY 2025 are estimates from RAND's Air Force Capability Investment Strategy (AFCIS) model (April 2002 version of the model using December 2001 data as input). According to AF/XP staff, updates to this dataset after December 2001 were considered to be classified for years 2002 and beyond.

RAND *MG348-1.3*

Nonetheless, Figure 1.3 shows that the greatest changes in the Air Force aircraft inventory of the future will be in fighter aircraft, with the introduction of new airframes with advanced capabilities.[12] The F/A-22, for example, will have supercruise capability, which means that it will be able to maintain speeds faster than the speed of sound without using the afterburner. It will also have vectored thrust, which allows the aircraft to maneuver at low speeds in ways that conventional aircraft cannot. Both the F/A-22 and the F-35 incorporate stealth technology, and both will be fly-by-wire with side control sticks similar to the F-16. Both aircraft will introduce advanced sen-

[12] Including UCAVs (unmanned combat aerial vehicles).

sor fusion technology, which, in the words of the Boeing Company (2001), means that "targeting, detection and tracking information is fused from multiple sensors to create a single input to the pilot."

The capabilities of new aircraft like the F/A-22 and the F-35 will put new demands on pilots that may require the development of new skills. While there are no current public plans to develop new bomber, tanker, or mobility aircraft, new information technology systems such as those used in the F/A-22 may be transferable to the older aircraft; this possibility and new missions may introduce new skills for nonfighter pilots as well. If it is determined that those new skills should be taught in undergraduate pilot training (as opposed to IFF or FTU), the capabilities of trainer aircraft might need to be changed. As a minimum, aircraft used in undergraduate pilot training must provide the foundation for these new skills to be developed. While these changes might have more of an impact on the training of fighter pilots, and hence on the decision to replace the T-38C, information technology changes are also important for mobility pilots, and could have an impact on their training and the decision to replace the T-1A.

Research Approach

Our first task was to describe the mix of operational flying environments to which pilot training graduates will transition in the future. To do so, we collected information about plans for future aircraft and airpower strategies and scenarios outlined in defense documents, and also analyzed research on airpower in the future. We spoke to Air Force leaders in positions of planning and decisionmaking in relevant areas, and interviewed pilots who recently served in operations in Iraq and Afghanistan to hear their impressions of future trends they saw emerging during their deployments.

Our second task was to develop a taxonomy of flying skills currently developed during different training phases. We originally planned to identify and classify the skill sets developed during different preoperational training phases to determine first whether the training in place is sufficient for today's technology and missions, and second the baseline for assessing whether this training will suffice for future aircraft and missions. To prepare to develop this taxonomy, we analyzed the current training syllabi for each of the trainers and operational aircraft and spoke with instructor pilots and recent graduates from the training programs. As we progressed in our research, we found that the responses to our interviews were so general in the area of skills development, though very specific in answer to our research question, that the development of a taxonomy of skills turned out not to be important.

Third, we focused on the training pipeline, particularly how well the training aircraft and skills taught in different phases prepared pilots for their operational aircraft, and the degree of adaptability required to move forward through the training stages. We also considered what the best stage is for teaching each skill. For example, some skills currently taught in later stages could be taught sooner because of improved simulator or instructional technology. To keep our research grounded in the practical, the feasible, and the most up-to-date circumstances, we conducted interviews with junior, senior, and instructor pilots and with trainees in each aircraft community. Our findings were considered in the context of the even greater leap that would have to be made transitioning from current trainer aircraft and instruction to future aircraft and battlefield demands.

Our fourth task was to take the information acquired in task one, compare it to the information collected in tasks two and three, and analyze the adequacy of the current training and current training aircraft for meeting the requirements of future aircraft and missions.

Methods for Pilot Interviews

Our sampling strategy was designed to match our research questions; thus, we sought to interview pilots, both instructors and trainees, in every phase of training and in most aircraft communities, with special attention to using their experiences to understand potential changes in skills required for pilots of newer aircraft like the F/A-22 and the F-35. Given the current deployment schedule and demands on pilots' time, at the bases we visited we requested interviews with one or two pilots at a time rather than with focus groups. Each interview session lasted from 30 to 45 minutes and was conducted by two researchers: one leading the interview and the other primarily dedicated to taking notes. Table 1.1 shows, by command, the units we visited and the number of pilots interviewed at each unit. Our intention was not to produce a large-scale random opinion survey of pilots, but rather to sample for maximum variation in order to capture a wide range of experiences for our analysis to consider.

The experiences of the pilots we interviewed covered virtually all major operational aircraft in the Air Force inventory. Pilots in our sample also included graduates of generalized UPT (who flew the T-37 and T-38), SUPT (graduates who had been tracked to either fighter/bomber training in the T-38 or mobility aircraft training in the T-1A) and JSUPT (pilots who had flown T-34 aircraft with the Navy and then tracked to either T-38s or T-1As at Vance AFB, and pilots who flew T-37s with the Air Force and then tracked to T-44 training with the Navy).[13]

Preview of Findings

Our primary finding is that in the context of SUPT, pilots agree that the T-38C and the T-1A are capable of providing student pilots the skills they will need to fly future aircraft through 2040. We were surprised that this feeling was essentially universal: only one or two pilots recommended the replacement of the T-38C, and their concerns were related more to the age of the aircraft than to the skills that can be taught in it. A recurring theme in our interviews was that cockpit demands on future pilots, such as information management and the ability to filter and prioritize information from many sources, will increase, but that basic flying skills currently taught in pilot training will always be necessary.

---

[13] Our interviews included two Royal Air Force exchange pilots from the United Kingdom as well.

Table 1.1
Interview Locations and Personnel Interviewed

Unit	Base	Sample Pilot Backgrounds	Number Interviewed
AETC			
97 AMW	Altus	C-141, C-5, C-17, KC-135	20
58 SOW	Kirtland	H/MC-130, CV-22, MH-53J, HH-60G, UH-1N	20
47 FTW	Laughlin	T-37, T-6, T-1, T-38	20
56 FW	Luke	F-16, KC-135, C-130	20
479 FTG	Moody	T-6, T-38C, F-117, F-15E, F-111, A-10	21
12 FTW	Randolph	T-37, T-6, T-1, T-38	28
43 FS	Tyndall	F/A-22	1
ACC			
2 BW	Barksdale	B-52	16
49 FW	Holloman	F-117, F-111, F-16, F-15	10
1 FW	Langley	F-15C, F-15E	10
ACC/DR-JSF	Langley	T-37, T-6	1
ACC/DR-F/A-22	Langley	A-10	1
57 WG	Nellis	F-16, F-15, F-117, KC-135, Predator	8
509 BW	Whiteman	B-1, B-2	20
AMC			
AMC/DO Staff	Scott	C-17, C-141, C-5, KC-10, C 130	11
AFSOC			
16 SOW	Hurlburt	AC-130, MC-130, HH-53, C-5	15
Other Organizations			
AF/XPX	Pentagon	N/A	3
ASC/YTG	Wright-Patterson	N/A	5

Situational awareness (SA) is a term that interviewees used to describe a unique and important trait of flying that is required by all pilots. It can be described as the ability of a student to assess him- or herself and the aircraft "in relation to the dynamic environment of flight, threats, and mission" and the ability to forecast events and decide what to do based on that assessment.[14] The majority of the pilots

[14] This definition of SA is paraphrased from the T-38A course training standards (Air Education and Training Command, 2002b, p. 23). In a Department of Transportation report, Uhlarik (2002) says that the most commonly cited definition of SA (which he calls *situation* awareness) is one suggested by M. R. Endsley (1995, p. 36): "Situation awareness is the perception of elements in the environment within a volume of time and space, the comprehension of their meaning, and the projection of their status in the near future."

we interviewed said that undergraduate pilot training should focus on developing this skill, and the T-38C and the T-1A (though perhaps to a lesser degree) were more than adequate for the task. At the same time, those we interviewed indicated that the use of the advanced technological tools of newer aircraft (e.g., for navigation and weapons deployment) could be taught in later phases of training. A student with well-developed SA could handle the new demands in the cockpit; a student without good SA would have more difficulty. In other words, exposure to advanced cockpit resources is less important in undergraduate pilot training than is the rigorous development of basic flying skills.

From fighter pilots, we heard concerns about graduates of SUPT/IFF going directly to the F/A-22 and the F-35. Both of these aircraft are single-seat fighters, and there are currently no plans to build two-seat trainer versions of the aircraft. This is important because these aircraft can pull more sustained g's in flight than can the T-38C, and the possibility of g-induced loss of consciousness (GLOC) is greater than in the aircraft flown in SUPT. GLOC is a real danger: from 1982 to 2002 there were 559 reported GLOC incidents in the Air Force. Twenty of them resulted in fatalities, and all of the fatalities occurred when a pilot was flying solo (Lyons et al., 2004). Current single-seat fighters like the F-15 and F-16 have two-seat trainer versions that are used in initial training, so new pilots in these aircraft learn to adjust to higher g-capability while flying with an instructor.[15] The F/A-22 is capable of sustaining similar g-forces, and fighter pilots we interviewed were concerned that graduates of SUPT would not be able to manage this capability without some type of preparatory training. They did not feel that SUPT was the appropriate phase to do this, but speculated that training between IFF and FTU (for example, in an F-16) would be necessary to expose an

[15] Graduates of SUPT are not assigned to the single-seat F-117; only pilots qualified as 4-ship lead in other fighter aircraft are allowed to fly it. This is because the F-117 flies missions as a single aircraft, and the Air Force feels that only mature, experienced pilots are qualified to do this. The A-10 is a single-seat aircraft, but it is not capable of sustaining as many g's as are the F-15 and F-16, and initial sorties are flown with an instructor in a chase aircraft.

SUPT graduate to a higher g environment with an instructor before training in the F/A-22.

Mobility pilots did not have comparable concerns about the physical demands of future aircraft and missions. They emphasized the importance of producing pilots with good situational awareness.

Organization of the Report

The rest of this report is organized as follows. Chapter Two begins with a short look at the history of undergraduate flying training in the Air Force and the justifications used for changing from generalized UPT to SUPT. It also includes perceptions of the current SUPT program shared by the pilots we interviewed. This provides insight into when, or if, students should be tracked and what skills a person needs before being awarded wings. Chapter Three describes the Air Force's view of future peace- and wartime missions and how these missions might affect the capabilities required of future aircraft and the skills pilots will need to fly them. It also introduces insights from the international flying community and the results of our pilot interviews. Chapter Four uses more information from the pilot interviews to address the adequacy of the T-38C and the T-1A for providing the skills needed in future aircraft. Chapter Five concludes by raising more issues that must be considered in order to decide to replace or retain current trainer aircraft. Chapter Six outlines further work that must be done to make an informed decision about replacing the T-38C and the T-1A.

CHAPTER TWO

Current Training

A Brief History of Undergraduate Pilot Training

The Early Years[1]

Over the years, the Air Force has employed different concepts in training pilots. For 20 years (encompassing World War II and Korea), the Army Air Corps and then the Air Force employed variations of a specialized undergraduate pilot training system. As an example, in 1952, a student pilot flew a T-33 or a B-25 in the advanced phase, depending on his follow-on aircraft.

A 53 percent attrition rate in 1950 triggered a reexamination of the flying program. A 1952 study identified "lack of motivation" in 28 percent of those who failed to graduate. In September 1952, a Project Tiger study team recommended that "all pilot training should be built around the assumption that each student was being trained to fly a jet fighter in combat." Within 10 years, the Air Force had transitioned to a generalized UPT system. All students flew the T-37 in the primary phase and the T-38 in the advanced phase.

Studies and More Studies

A number of studies were accomplished in the 1960s, 1970s, and 1980s comparing generalized to specialized UPT. Additionally, the

[1] Information in this paragraph is from Emmons (1991, pp. 1–8).

expanding conflict in Southeast Asia and projections of high pilot production goals raised issues about the need for a new trainer aircraft. Thus, the recommendations of these studies must be understood in the context of how their underlying assumptions were influenced by the world military situation.

In 1965, an Air University Study concluded that the then-current training philosophy and practices would generally remain valid (Shircliffe, 1975, pp. 1–2). Two years later, a 1967 Air Training Command (ATC)[2] Study recommended the following:

1. Air Force Systems Command (AFSC) should develop a more accurate measure of motivation and the required traits for selection of pilot candidates.
2. Headquarters (HQ) U.S. Air Force (USAF) should reestablish an aviation cadet program.
3. The Air Force should conduct a study on pilot career management.
4. The Air Force should conduct a study on using a single aircraft for UPT.
5. The Air Force should retain the T-41 flight screening program (Shircliffe, 1975, p. 2).[3]

As can be seen from the fourth study conclusion, ATC was already considering an airframe change even in the 1960s.

1969–1972 Mission Analysis

In January 1969, ATC initiated a mission analysis on the future of undergraduate pilot training. This huge undertaking took three years to accomplish and involved more than 70 people. There were three broad reasons for the mission analysis (Mission Analysis Study Group, 1972). First, there were concerns about equipment deficiencies because of the high UPT production requirements (although re-

---

[2] ATC became AETC on July 1, 1993.

[3] The T-41 is a version of a Cessna single-engine propeller aircraft.

cent decreases in projected production had extended the life of the fleet). Second, ATC recognized that there had been rapid advances in flight simulation, and this raised the question of why advanced simulators were not being used in UPT. Finally, there had been few substantive changes in the pilot training process, and while the training was very efficient, it was thought that new learning theories might offer breakthroughs in how pilots are trained.

The study team examined selected representative aircraft for the 1975–1990 timeframe to develop future training requirements. They developed eight different groupings of common tasks, some of which overlapped (Mission Analysis Study Group, 1972, p. 11). Specifically, the tasks were those common to the following:

- nearly all operational aircraft
- most operational aircraft (the study showed no distinction between this category and the "nearly all" category)
- air-to-air, air-to-ground, reconnaissance, and forward air control aircraft
- air superiority and intercept aircraft
- close air support, interdiction, forward air control, and reconnaissance aircraft
- strategic bombing, transport, refueling, and rescue aircraft
- strategic bombing aircraft
- assault and intratheater airlift, refueling, and rescue aircraft.

The study team identified current and planned future aircraft systems for the analysis. One future aircraft was the so-called AMI, an advanced manned interceptor, as a replacement for the F-106A. Also postulated was a vertical takeoff and landing (VTOL) fighter for air superiority. These systems, as well as then-current aircraft in the fleet, were selected for the analysis—though in hindsight, none of the postulated future aircraft was ever produced. This highlights one of the obvious difficulties of building systems to meet future needs: one never really knows what the future holds.

The team visited nine airbases and evaluated aircraft tasks in the above eight areas. Additionally, it evaluated UPT and SUPT against various learning theories (Mission Analysis Study Group, 1972, pp. 33–49), and looked at the potential for increased use of simulators and other future training media (Mission Analysis Study Group, 1972, pp. 50–71).

The study team concluded with a number of findings; five of the most significant follow (Mission Analysis Study Group, 1972, pp. 114–115):

1. Future UPT will require higher-quality training.
2. The number of training requirements determines if a program should be generalized or specialized. The study team identified 30 training requirements that included 10 new requirements that were not taught in 1972 (see Figure 2.1). The study team felt that a specialized system would be required if all 30 requirements were to be taught in undergraduate training.
3. A 10 percent attrition level in UPT is a realistic goal with centralized selection and if training improvements are implemented.
4. Flight simulation will provide the breakthrough for increased training quality at lower costs. The study group recommended simulation for instrument training by 1976 and full mission simulation by 1983. The study team actually traded flying hours for simulation hours in a number of options.
5. Current (as of 1972) trainer aircraft were adequate for the future. The study team recommended the purchase of more T-37s and advised that both the T-37 and T-38 would require avionics add-ons to operate in the future UPT environment.

The steering committee to which the study group reported accepted most of the recommendations, but rejected the purchase of additional T-37s. Without explanation, the steering committee deferred the decision to procure new conceptual aircraft until the 1979–1982 time frame.

Direction Changes

As ATC began installing new procedural trainers at the main bases, the question of SUPT versus UPT again arose. Lt Gen John W. Roberts, Air Force Deputy Chief of Staff for Personnel (HQ AF/DP), stated in September 1974 that "the Air Force goal has been to produce a universally assignable pilot from UPT; however, today's budgetary constraints may dictate that we change that policy. The logical result of such a policy change may be some type of a 'two-track' pilot training system" (Emmons, 1991, p. 12).

A 1976 ATC study that compared UPT to SUPT concluded ". . . the purchase of new aircraft to support specialized training cannot be justified in view of today's austere budget, programmed low UPT production and the resulting aircraft fleet-life extension this affords, and MAJCOM [major command] acceptance of the current, high-quality UPT graduate" (Emmons, 1991, p. 13). Based on this study, ATC recommended that the Air Force retain the generalized pilot training that produces a universally assignable pilot (Emmons, 1991, p. 14).

SUPT, however, developed a life of its own, and despite numerous ATC rejections and additional ATC studies, Gen Roberts (who by then was the ATC Commander [ATC/CC]) in 1977 wrote, ". . . the only training system that can optimize both quality and cost is a specialized training system" (Emmons, 1991, p. 14). The Strategic Air Command (SAC) and the Tactical Air Command (TAC) both expressed opposition to the idea. The Air Staff supported the switch to SUPT and the plan to replace both of the current trainer aircraft (Emmons, 1991, p. 15). The Air Staff separated the T-37 replacement from the overall plan in order to expedite approval. In June 1979, the Defense Department approved the operational requirement document.

ATC actually selected an aircraft, the T-46, to replace the T-37, but tighter congressional funding limits and development problems with the aircraft killed the program. ATC then began a SLEP[4] for the T-37 that extended its service life to 30,000 hours and delayed the need for a replacement until 1999 (Emmons, 1991, p. 21).

---

[4] Emmons uses SLEP to mean *structural*, rather than *service*, life extension program.

Numerous starts and stops prevented ATC from actually procuring aircraft for an SUPT program for either phase until 1990 when the Air Force purchased the T-1 Jayhawk off the shelf (Emmons, 1991, pp. 14–53). In October 2001, the T-6 Texan also started replacing the T-37.

In an AETC History Office monograph, one author concludes: "It had taken ATC a long time to come this far. It was just over 31 years since the command had dropped specialized undergraduate training in favor of generalized training. And it was almost 13 years since the day in March 1977 when General Roberts, the ATC commander, had advocated a return to SUPT" (Emmons, 1991, p. 55). Figure 2.1 highlights this history by showing some of the aircraft changes in pilot training since 1947 and how they will extend through 2020. The number of aircraft used before 1960 and the long tenure of the T-37 and the T-38 in generalized UPT are notable.[5]

Observations

The Air Force has not been particularly successful in finding money to fund the purchase of new training aircraft, and has had to argue that new aircraft will save training costs. For example, the T-1A flying-hour cost is one-third the cost of the T-38. Likewise, the T-6 is considerably cheaper to operate than the T-37 (U.S. Air Force, 1994, Attachment A2-1). While there was a push for SUPT for many years, its implementation was more the result of finding a way to ensure the survival of the T-38 by reducing flying hours with the incorporation of the T-1A. It would seem that without major changes in the training requirements of future aircraft, any prospect of replacing the T-38 will likely be driven by lower operating costs.

---

[5] Figure 2.1 does not include all aircraft that were used prior to 1959. It is meant to be illustrative of the change from specialized training before 1960 (when, for example, students flew either B-25s or T-33s, depending on their aircraft assignment) to generalized UPT, and the return to SUPT in 1992. The phases in this chart are consistent with Figure 1.1, but there have been confusing changes in terminology over the years. The T-38 phase is sometimes referred to as the basic phase. Prior to 1961, the analogue of the T-37 phase was called the basic phase.

Figure 2.1
Aircraft Used in Air Force Pilot Training

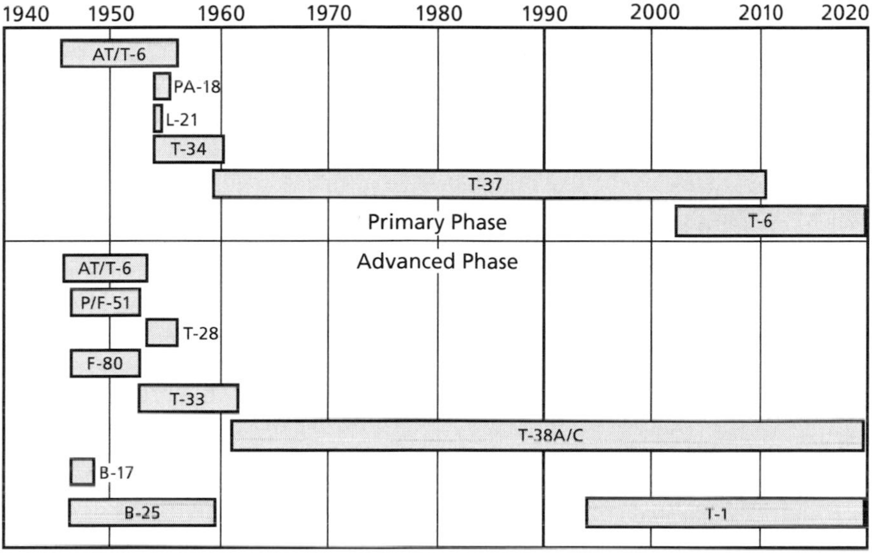

SOURCE: Information derived from Air Education and Training Command (2002a).
RAND *MG340-2.1*

Pilot Opinions of Current SUPT

Among those we interviewed, there was overall satisfaction with the pilots produced under the current SUPT program. However, the majority of those to whom we talked felt that the tracking decision is made too early. We describe below some specific comments by the flying community.

Fighter Pilots

Instructor pilots and fighter pilots agree that SUPT provides an appropriate foundation for IFF and FTU training. In addition, the avionics and engine improvements introduced by the T-38C have made it an excellent transition aircraft. Those we interviewed felt that using the T-38C in SUPT would bring more improvements to the performance of students selected for fighter and bomber aircraft.

Most fighter pilots we interviewed felt that the tracking decision was made too early. This opinion was expressed not only by pilots who had graduated from generalized UPT, but also by some who were graduates of SUPT. Many of the generalized UPT graduates related stories of colleagues who had not performed well in the primary phase of training, but had blossomed in the T-38 phase. They also told us of fellow students who had entered UPT with extensive commercial aviation experience and did very well in the T-37, but could not adjust to the high-speed T-38, and went from being at the top of the class in T-37s to the bottom of the class in T-38s (if they were not eliminated). The first case was used to argue that potential fighter pilots could be wrongly tracked to mobility aircraft in SUPT and the second to argue that SUPT might be sending people to the fighter track when they should not be there.[6] Both situations were used to argue for making the tracking decision later in SUPT, after students have had more flying time, and perhaps after all students have had some experience in a fast aircraft like the T-38.

It is difficult to assess this anecdotal evidence. Since students are tracked after the T-37/T-6 phase, it is impossible to know if T-1A students who did poorly in the primary phase might have done well in the T-38 if they had been given a chance. It would be possible to examine the other case of incorrect tracking by checking statistics of students who performed well in the T-37 but did poorly in the T-38, but our resources did not allow a detailed analysis. However, 1st Lt Kim Hoss described such an approach in a briefing entitled "A-10 FEB Review" produced by the AETC Studies and Analysis Squadron on August 27, 2003. The briefing analyzed seven SUPT graduates who met Flying Elimination Boards (FEBs) because of poor performance in A-10 training. In this study, six of the seven students in the study had been in the top half of their T-37 class, and four of these six ended up in the bottom of their T-38 class. A finding in another

[6] It was also used to argue that some people who fail in the fighter track might have succeeded if they had been correctly tracked to the nonfighter course.

study (Hoss, 2002) was that of the bottom 50 percent in a T-37 class, 12 percent finished in the top half in T-38s. These two reports show that there is some support for the anecdotes we heard in our interviews about reversals of class standing from T-37s to T-38s.

Another complaint by some fighter pilot graduates of SUPT we interviewed was that nonfighter pilots dominate the instructor-pilot pool in the primary phase of training, and that this can discourage capable students from stating a preference for the fighter or bomber track in training. One F-15 pilot we interviewed said that his T-37 class was divided into two flights, one commanded by a fighter pilot, and the other commanded by a nonfighter pilot. All of the students in the fighter pilot's flight ended up in the fighter track, and all of the students in the nonfighter flight ended up in the nonfighter track. We did not attempt to verify this recollection, but it is representative of attitudes we heard from other fighter pilots. We also heard that the fighter pilot track in SUPT is considered to be riskier than the T-1A track, and some students with the skills to fly the T-38 request the T-1A track because they think chances of graduation are higher. Again, we have no statistics to back up this claim, but several pilots we interviewed expressed this opinion.[7]

Mobility Pilots

Mobility pilots we interviewed also agreed that graduates of SUPT were well prepared for their assignments to transport and tanker aircraft. The T-1A is an excellent medium for training students in cockpit/crew resource management (CRM), and for exposing them to tools that they will see in operational aircraft, such as weather radar, autopilot, and flight management systems.

We did hear some concern that SUPT focuses on producing good copilots, and that graduates from the T-1A tend to be less independent and less willing to make decisions on their own than had UPT graduates in the past. This type of comment was often paired

[7] The feeling is not universal. One F-16 student pilot we interviewed said that his memory of SUPT was that the T-1A students worked harder than the students in the fighter track because they had to spend so much time on flight planning.

with an opinion that graduates of generalized UPT had more confidence in their decisionmaking because of their experience in the T-38. In remarks that mirror the fighter pilot opinion that the mobility track in SUPT is considered less demanding than the fighter track, some mobility pilots noted that some in the Air Force consider T-1A graduates somehow less capable.

Finally, and to us surprisingly, many mobility pilots agreed with their fighter counterparts that the SUPT tracking decision is made too early. They felt that students could make better choices about what to fly if they were given opportunities to have more experience.[8]

Bomber Pilots

While bomber pilots generally validated the SUPT program, comments varied on the best approach to tracking. While the T-1A has CRM training advantages that might be useful for the B-52 with its crew of five, pilots felt that the T-38 track was good preparation as well, since it teaches communication by radio and much of the B-52 crew coordination is done by intercom. Because the flight characteristics of the T-38 are similar to the B-1, B-1 pilots felt that the fighter track in SUPT was good preparation for them. We talked to B-2 pilots from a variety of backgrounds, and they, like the fighter and mobility pilots we interviewed, tended to feel that track selection in SUPT was too early. A common comment was that some experience in the T-38 would help both students and instructors make better decisions in track selection.[9]

Special Operations Forces (SOF) Pilots

The SOF community, with missions such as infiltration, combat search and rescue, resupply of special operations forces, close air sup-

---

[8] Student pilots selected for C-130 aircraft used to go through the mobility track in the T-1A. They now are tracked to training with the Navy in the T-44 aircraft—a twin-engine turboprop. Almost all of the pilots we talked to agreed that the T-44 training was better preparation for the C-130.

[9] Bomber pilots also noted that since they fly a limited number of sorties per month, some type of aircraft that allows them to fly between bomber sorties is necessary for them to maintain proficiency.

port, and air interdiction, and aircraft such as UH-1N and MH-53J Pave Low helicopters, A/H/MC-130s, and the CV-22, has diverse requirements, but we heard few complaints about the preparation provided by SUPT. Crew management and good situational awareness are extremely important for all of these missions. We heard that, by the nature of the SOF environment, most Air Force Special Operations Command (AFSOC) trainees are initially overwhelmed by cockpit demands.

Conclusion

Overall, the pilots we interviewed were satisfied with the SUPT graduates with whom they trained or flew, though there was a prevailing opinion that the tracking decision is made too soon.[10] Before discussing their opinions about the potential need for new trainer aircraft in the future, we must discuss the missions that the Air Force plans to undertake and the skills that these missions will require.

---

[10] This feeling is not universal, of course. Some fighter pilots were adamant that SUPT was a failure, and that the Air Force should return to a single-track pilot training approach so that graduates are universally assignable. On the other hand, one colonel we interviewed noted that there are likely individuals for whom SUPT is unfair in that the tracking occurs too early for their skills to be recognized, but on average, the SUPT graduate is better for the Air Force because SUPT starts training students earlier for the missions they will be flying operationally.

CHAPTER THREE

Future Air Force Missions and Projected Pilot Skills

> There is no denying that the world is a decidedly different place than the one we knew in the previous century of world wars and our struggle against communism. Today, our adversaries' goals include creating terror through disruption of our economic system and by striking American interests at home and abroad. . . . More troubling, in this new era of stateless actors, these aggressors may be non-deterrable; at least by the traditional means we have employed to prevent wars among nations.
>
> —*Roche, 2002*

The Future in Air Force Planning Documents

The future is uncertain, and with uncertainty comes the need for the Air Force to prepare for the possibility of a broader and different portfolio of missions than it has in the past.[1] Preparing for different missions could likewise imply the need for a different type, amount, and mix of major weapon systems (MWS) and tactics than exist to-

[1] See Ochmanek's work (2003) for a discussion of the type of changes that the Air Force will see in the post–9/11 environment. Another historical set of readings with relevant impacts on change can be found within the series *New World Vistas: Air and Space Power for the 21st Century* (U.S. Air Force Scientific Advisory Board, 1995).

day. This may also require greater flexibility than SUPT provides in assigning graduates to a variety of Air Force systems.

In the context of this study of SUPT, we expect that changes to future systems and missions will affect the type of training aircraft systems that should be procured and the type of training that will be undertaken. Today's pilot training methods and training aircraft should be structured and procured in such a way that they will support the anticipated operations of tomorrow. In other words, today's training pipeline should be primed in order to be flexible and prepare for tomorrow's demand. Although we place significant caveats with respect to our limited abilities in actually predicting what the future may hold for pilots, we still believe that it is important to understand the ideas and plans that are in the works.

In this chapter, we explore two specific areas of information. The first area examines unclassified Air Force strategic plans as well as published material on the nature of future operations and aircraft. The second discusses themes that our project team found during the interviews with pilots across the Air Force. Analyzed together, these two streams of information will provide the context of the demand for future competencies and training aircraft systems.

Review of Air Force Literature and Data

We began our study of future requirements by examining relevant material that has been written on the subject of the future of the Air Force, with a specific focus on two components: activities affiliated with flying operations, and key characteristics of the major combat weapon systems that will be used.[2] This material was derived from three subcategories of information: official Air Force policy, Air Force planning data, and nonofficial literature written on the subject.[3] All sources used for this analysis are unclassified.[4]

---

[2] The phrase *combat weapon systems* refers, in this context, to fighter, mobility, tanker, and special operations forces aircraft. This description excludes training aircraft, by definition.

[3] Sources for this latter category included material published in conjunction with the professional military education (PME) schools at Air University, the *Airpower Journal*, the *Air*

In general, material written within the recent past indicates a broad consensus that the nation, and more specifically the Air Force, will face a changing threat environment.[5] This is especially true in light of the attacks that occurred on September 11, 2001. As Secretary of the Air Force James G. Roche articulated in a speech at Air University in 2003, "this is a new age of warfare" (Roche, 2003).

Although virtually everyone with an opinion about the future of the Air Force would agree that change is likely (we could not find literature that was counter to this), there is less consensus on what the future will bring and the speed and scope of change with respect to pilots and flying operations. Some authors think the next 100 years of airpower will be even more revolutionary than the past century of flight (Link, 2001), while others predict more gradual changes to the landscape. Synthesizing the vast amount of material that is written on the subject, we do not see these predictions as orthogonal perspectives as much as different opinions about the degree of change and the amount of time that will pass (e.g., writers see relatively more change as we look further into the future).

For purposes of this review and our study, we have narrowed our field of vision to include the present (2005) to the 2025 timeframe. This window of time is consistent with the time horizon over which the next procurement of trainer aircraft would potentially occur—and is also consistent with our belief that projections beyond 20 years or so are highly suspect because of uncertainty and the inherent complexity of the situation.[6]

Force Times, speeches by senior Air Force leaders in public forums, and work from our institution, the RAND Corporation.

[4] The authors make no claim on the accuracy or correlation of this unclassified information with official, Air Force classified planning documents. We did not review any classified material (data, planning documents, or programming/budget information).

[5] Two recent works provide significant insight on these points (Lambeth, 2000; Khalilzad and Shapiro, 2002).

[6] Our colleague, Paul Davis (2002), has written on the same subject and suggests methods for dealing with such uncertainty in defense planning.

In general, we observed that the future of the Air Force can be characterized by the following two statements:

- Air Force flying operations will occur around the clock and in all weather and geographical conditions in a joint warfighting environment, and will place increasing demands upon pilots to integrate information from across a myriad of sources.
- Air Force major combat weapon systems will be stealthier, employed at higher speeds, rely more upon information technology, and be more lethal than the systems of today.

The remainder of this section describes the flying operations and future weapon systems in greater detail.

Changes in Flying Operations

Within the next two decades, we do not expect flying operations to be markedly different than they are today. Even with the expectation of increasing reliance upon unmanned aerial vehicles,[7] the literature and policy suggest that pilots will be a key component in the operation of aircraft (including UAVs) (Jefferson, 2000) using the same flying skills that they have in the past. Principles of airmanship are not expected to change significantly with the exception that pilots will be required to assimilate more sources and amounts of information (Secretary of the Air Force, 2002, p. 139). Where change is expected to occur, it is likely that this will surface in the greater frequency of operational activities that pilots have begun to experience during the past decade. The following six themes were distilled from a survey of literature focused on Air Force flying operations—these six represent trends in flying operations that have been observed over this time frame.

1. Flying operations will occur around the clock in all weather and geographical conditions. With the advent of better navigation systems and technology such as night-vision goggles (NVGs), the Air

[7] George Cahlink's piece on "War of Machines" (2004) is a good representative of current themes that are focused on the use of UAVs in today's and tomorrow's flying environment.

Force has become increasingly dependent upon night operations. Development of these technologies has affected the temporal aspects of how the Air Force can conduct operations—to such an extent that by 2025, flying in all operational conditions, around the clock, will more likely be the norm than not (Secretary of the Air Force, 2002, p. 143). As an Air Force Research Laboratory (AFRL) Report indicated:

> Night operations have become the mainstay of air operations in combat and continue to increase in importance. Every aircraft operates at night and most operate with NVGs during combat and combat training. The tactical advantages gained by operating at night with the aid of night vision technology, especially NVGs, have been well documented. In order to fully exploit the advantages gained by night operations, our force needs to be well trained and equipped to operate at night. The earlier stages of a pilot's training have been and still tend to be oriented towards daylight operations. USAF senior officers believe the introduction of effective nighttime NVG-aided training and infrastructure has lagged behind the operational need of a 24-hour capability (Martin, 2004, p. 1).

The latter part of this statement highlights an important contrast between how pilots are initially trained today compared to what the operational flying environment requires and will continue to require in the future. For the most part, the SUPT environment is focused upon daylight operations while the operational world is increasingly focused upon flying at night.

2. Flying operations will require much shorter response times for bringing airpower to bear upon the enemy. Technology has enabled the Air Force to shorten the response time with respect to how fast targets can be engaged in the combat environment. For example, during the first night of Gulf War II, a lone B-1B bomber carried out a massive strike on what the coalition described as a "leadership target" (Sanger and Schmitt, 2003). Although later study of the strike indicated that Saddam Hussein was not killed in the attack, the operation highlighted that airborne strike aircraft could receive real-time intelligence information and act upon such information within a relatively short time frame (15 minutes) (Gatlin, 2003).

In a 2001 briefing on Air Force transformation, Maj Gen Dave Deptula highlighted several points related to the future combat environment. One of his key points was the potential for "future aggressors with asymmetric capabilities" to attack—something later borne out on September 11. In Gen Deptula's analysis, the Air Force would be required in the future to respond to such attacks in relatively short periods of time—without much opportunity for planning (Deptula, 2001).

Recent Air Force research and writing on the subject has also called for more focus to be placed on the doctrinal aspects of how time-critical targeting will be conducted in future warfare.[8] The use of space, ground, and air-based sensors will intensify the type and amount of information that a pilot will have to consider in making targeting decisions.

3. There will be a desire to minimize the loss of pilots and manned aircraft while simultaneously increasing mission effectiveness. Weapons like the Joint Direct Attack Munition (JDAM), Joint Stand-Off Weapon (JSOW) and other precision-guided munitions (PGMs), along with more robust cruise missile variants, will allow combat operators to stay further away from hostile environments during attack.[9] After the first Gulf War in 1991, the Air Force purposefully sought to develop and procure PGMs to achieve greater accuracy while simultaneously allowing for a greater distance between pilot and threat. The impact of such precision has been measured in terms of reduced sorties and lower residual casualties among nonwarfighters. During the last decade, it has been estimated that PGMs have increased destructive power over a thousandfold relative to older, non-PGM systems (Lambeth, 1996). More capable and further-reaching air-to-air munitions coupled with higher resolution ra-

---

[8] See Deale, 1999, and Grant, 2003, for contemporary examples on how Air Force Doctrine is changing to meet real-time needs.

[9] For a comprehensive history of PGMs and the type of benefits that can be realized from their use relative to non-PGM weapons, see (former U.S. Air Force Historian) Richard P. Hallion's work, "Precision Guided Munitions: The New Era of Warfare" (1996).

dar will also allow fighter pilots to reach beyond the horizon to engage the enemy (Ennett, 1999, p. 34).

In the most extreme case of minimizing threats to life and systems, the Air Force will employ unmanned combat aerial vehicles (UCAVs) to conduct combat, surveillance, and intelligence-gathering operations. As Air Force Chief of Staff General Jumper stated in 2003 to the Air Warfare Symposium,

> We are going to take this whole notion of unmanned aerial vehicles (UAVs) and remotely piloted vehicles and change the name of remotely piloted aircraft to RPA, to fully capture the kind of things that you are doing in something like the Predator, where a pilot is required and pilot actions are necessary to take the responsibility for dropping weapons and putting aircraft on targets; the same level of responsibility, we feel, as in piloted aircraft. And then the UAV name can be reserved for things that do not quite require so much of a human interface, such as the Global Hawk.

> We have to get it right, on this notion of remotely piloted aircraft and UAVs. One such issue we are dealing with is the issue of the unmanned combat aerial vehicle (UCAV), the conventional armed unmanned vehicle. What we have to get right is that we have to make sure that we fully understand what those leveraging qualities of unmanned aircraft are. And that we are not going out to buy something merely for the novelty of taking the person out of the aircraft. The thing that makes a Predator so leveraging for us is the fact that it stays airborne for 24 hours. It has persistence. It has endurance. It does things that a person could not do in that airplane. The same thing with Global Hawk. So if we are going to take advantage of those same qualities in an armed vehicle, then we should demand an order of magnitude increase in the capability of the vehicle that we go out and buy. And we have to look very carefully and be very cautious of going out and getting something that does not advance the mission and is only attractive because of the novelty of not having a person in it (Jumper, 2003).

4. Other types of flying operations (mobility and special operations forces, specifically) will require service members to operate in closer proximity to the enemy. Whereas fighter aircraft may not be required to operate as deeply into enemy territory relative to military engagements in the past, the future will likely require that mobility and special operations forces aircraft and crews operate in relatively higher threat environments than what was previously encountered. As the Chief of Staff of the Air Force (CSAF) said in 2003,

> In the area of global mobility, we are working on a concept of operations that takes us from the initial phases of rapid deployment, entry of data into the joint deployment, the loading of aircraft, the enroute visibility of what is on the aircraft, and the changing of the missions while enroute, to the ability to go from a concrete runway in the United States to a dirt runway somewhere in the middle of a contingency area, with all the information that has to pass enroute to make that safe (Jumper, 2003).

Compared to just a few years ago, the ability of aircrews to fly into hostile environments has changed the mobility mission. Recent C-17 operations in Iraq and Afghanistan are indicative of the capabilities that mobility brings, literally, to the fight. It is not uncommon today for mobility crews to take on hostile fire during routine operations. In this light, mobility pilots will be required to know tactical operations and flying techniques (defensive—and possibly even offensive) to thwart attack. The current environment is literally redefining the concept and definition of Combat Air Force (CAF) to include not just fighter aircraft, but mobility aircraft as well.

As recently as August 2004, Air Mobility Command has required that all of its aircrews study and adhere to Air Force tactics, techniques, and procedures that are outlined in a series of training publications called AFTTP 3-3 (Air Force Tactics, Techniques, and Procedures 3-3). The mobility version of AFTTP 3-3 details tactics and techniques for wartime employment. As the Vice Commander of AMC stated, "The Air Mobility Command's aircrews and weapons systems are more and more in harm's way—so the logical thing was to compile wartime tactics, techniques and procedures into a manual"

(Fazzini, 2004). Given the propensity for increased operations in these types of theaters, it is likely that crews will encounter hostile fire on a more routine basis.

5. Pilots will be required to integrate disparate sources of information (land, air, and space based) in real-time conditions. Operations will require the integration of airborne (manned and unmanned), space, ground, and advanced command and control systems. The ability to disseminate and act on information in a near real-time environment will be required to drive combat effectiveness in the future (U.S. Air Force, 2004c, p. 21). As former Vice Chief of Staff General Thomas S. Moorman stated in 1996,

> The 21st Century is upon us and I believe the trends I've spoken about will probably become realities. An integrated air and space program that combines total battlefield awareness and knowledge with rapid and dependable communications to get information to the decision maker or shooter, fully integrated with highly capable, survivable manned aircraft and a fleet of unmanned aerial vehicles (both with precision munitions) is the wave of the future . . . this capability merges the third and fourth dimensions of warfare, will be augmented by that fifth dimension, information.
>
> I believe that these new capabilities promise to usher in a new century that, if you will forgive a bit of parochialism, may very well be known as the Aerospace Century. Much as the Roman age was defined by the legions that conquered the known world, and the European Age of Discovery and Exploration was dominated by great naval fleets that secured trade and commerce well into the modern era, the 21st Century could well become the age of air and space power. Air and space power in the hands of democratic nations will be used to help secure the peace, provide humanitarian assistance and deter aggression throughout the world (Moorman, 1996).

Intelligence, surveillance, and reconnaissance, along with the ability to take information cues from sensors, will be key to successful operations. This type of activity will require pilots to filter, synthesize, and prioritize relatively large amounts of information in order to achieve mission success.

6. Operations will be more joint. The Goldwater-Nichols Act of 1986 mandated that the services work with one another more fully in order to achieve synergy and operational success. During the past two decades, there have been times when these relationships have functioned well; in other cases, not so well. One of the key lessons that the Air Force has learned in the current conflicts in Iraq and Afghanistan is the need for integration with the other services to achieve the mission (U.S. Air Force, 2004c, p. 20). As the Chief of Staff of the Air Force said in 2003,

> Working with the other services is very important. As the Army starts to contemplate its next generation of concepts of operations, we have to be mindful that the brigade combat team concept calls for troops deep behind enemy lines. Which means that our mobility forces are going to have to be able to penetrate, they are going to have to include things like precision air drop and air land in remote areas, things that now have to go into the global mobility concepts of operations. When you look at that global mobility concept of operations and you put those airplanes in that position, now they have to be fully cognizant of the total threat picture and the total common operating picture just as a fighter or a bomber aircraft would. And when you study the concepts of operations and you see the similarities between and among the concept of operations, it quickly leads to the conclusion that the thing you buy for one ought to be installed on all, rather than having a mobility team create their own situation awareness device. It also leads you to understand things that have to be further developed, like the need for precision airdrop (Jumper, 2003).

Air Force pilots will be required to interface more with the other services in this regard. The statement by General Jumper above focused on the mobility case, but fighter pilots and special operations pilots are just as likely to have to work with the information and communication infrastructure to ensure that targets are capably destroyed. There are two case studies (one successful and one not as successful) that highlighted this point.

In the case of Operation Anaconda in Afghanistan, the Army and Air Force did not coordinate close air support (CAS) operations well; the result was a loss of American lives that could have been prevented through better joint planning and execution. Contrast this with the success that the American forces had during the early days of the invasion of Iraq in which battle plans were coordinated and executed effectively between the air- and land-component commanders (Roche, 2003). Given the current U.S. focus on the war on terror, the demand for joint operations will continue to exist in the future and will be a key cornerstone of all operations (U.S. Air Force, 2004d).

Changes in Weapons Systems

Figure 1.3 displays potential changes in the Air Force's aircraft inventory by 2025, but it does not show one of the consequences of improved technology: decreases in the number of aircrew members required. Information technology and advanced navigation systems will automate information processing in ways that will substitute for personnel. This trend is occurring today in bomber aircraft: the oldest B-52 requires a crew of five compared to a crew of four in the B-1 and two in the B-2. Similar decreases are observed in cargo aircraft: a C-17 requires two pilots and a loadmaster compared to a C-141's crew of six (two pilots, two flight engineers, and two loadmasters).

The changes in aircraft inventory also do not show other technology influences. As the 2004 Air Force Posture Statement indicates, the Air Force Competency of Precision Engagement requires more precision intelligence (to be gathered through the operation of UAVs) and increased use of precision weapons (global positioning system [GPS]-guided) (U.S. Air Force, 2004c, pp. 21 and 22). This focus upon precision will likely extend into the future because of the desire for higher probability of target kills per weapon and the mitigation of collateral damage. For the most part, implementation of these weapons will require pilots to master advanced information systems and be able to program targets in flight. Another emphasis area and trend in future weaponry will be the increase in lethality of both air-to-ground and air-to-air munitions. An example of this trend is seen in the recent development of the small-diameter bomb (SDB). The SDB is a

250-pound weapon (roughly half the size of the smallest weapon in the Air Force inventory, the 500-pound MK-82) that may have lethality characteristics of much larger weaponry (Jumper, 2003). Because this weapon is much smaller and just as capable as larger analogs, a pilot will be able to carry considerably more firepower into combat than in the past.

Combat pilots will thus face increasingly complex technology and task environments. This will be driven by a more complex information management environment and by the need for higher proficiency in managing aircraft technology compared to today (e.g., more systems at faster speeds resulting in more efficient task prioritization). Together, these variables and other operational settings (integration of information sources, use of NVGs, and flying in all possible weather and geographical settings) are likely to put significant stress on the pilot's ability to maintain a high degree of situational awareness.

Although this discussion is focused primarily on fighter pilots, Air Force senior leaders envision that other support aircraft (such as tankers, for example) will see changes in operational focus due to advances in technology. During a speech given in 2003, General Jumper stated that the Air Force "will never again buy a single mission aircraft or platform . . . the idea of a smart tanker is to have these aircraft that always orbit very close to enemy lines, turn them into an IP address in the sky, and use them to pass information just as a computer network would around the battlespace for target information and other vital command and control information" (Jumper, 2003). It is clear from this discussion that changes in aircraft technology will influence the types of demands placed upon pilots, which in turn affect the type of systems and training that the Air Force must consider today.

An International View

Projecting skill requirements for Air Force fighter pilots around the world was one of the main topics discussed at a conference on fighter training that was held in London on June 16 and 17, 2004. The conference was sponsored by SMi Group, a London-based company that describes itself as a world leader in providing business-to-business information. Presenters at the conference included military leaders from several nations who are involved in training fighter pilots (though the Royal Air Force [RAF] dominated) and human factors/training experts and engineers from industries that develop systems to train fighter pilots. The audience was made up of military leaders, training industry representatives, and others interested in the question of how to develop effective fighter training systems to support current and future fighter missions. All speakers presented detailed insights into various issues involved in the production of fighter pilots with the skills necessary to meet current and future mission requirements.[10]

The consensus at the conference was that for future fighter pilots, information-processing skills will be more important than motor skills, and situational awareness will be more important than airmanship. The new fighter jets under development in the United States and in Europe will be easy to fly but difficult to manage, so pilots will have to be trained to do quite different tasks than they were trained to do during the Cold War. For example, instead of loading target coordinates before takeoff, pilots can now be tasked in the air to hit a target. Preparation for doing online targeting and tasking is an information-processing task that must be part of future training. Also, integration of UAVs into future missions will further increase the information-processing demands on pilots.

In the opinion of most European attendees, old-generation skills such as airmanship (meaning flying skills), though still relevant, will be less important in new jets, which will feature digital cockpits that require situational and tactical awareness capabilities to manage the information available. Information management in the new digital cockpits will be more complex, with far more sources of information,

---

[10] Research team member Richard Marken, a human factors psychologist, attended the conference.

and with a higher volume than exists in current cockpits.[11] Situational awareness is needed to guide prioritization tasks to determine which of these sources of information should be dealt with first. The supersonic cruise speed of some new jets will compress task time and make this prioritization process even more difficult.

The opinions expressed at the conference are consistent with the attitudes that informed the decision process that the Canadian Forces used when their Air Force was modifying its pilot training program in the late 1990s. Until the year 2000, students in the Canadian analogue of Phase I training flew an aircraft similar to the T-37 (the Tudor) and those selected to be fighter pilots flew the F-5 (a combat version of the T-38) in a follow-on phase. In 2000, the Tudor was replaced by the T-6 and the British-built BAE Hawk 115 replaced the F-5. When searching for a replacement for the F-5, Canadian trainers were interested in an aircraft that was similar to fighter aircraft flown in Canada and in Europe.[12] They wanted something with reasonable air-to-air capability, head-up display (HUD), and HOTAS (hands-on throttle and stick—this term refers to the placement of numerous switches on the stick and throttle) in order to help students develop cockpit management skills. Improved aircraft performance characteristics such as high angle of attack (AOA) and sustained high-g turn capability were not considered as important for student training as was the capability to improve cockpit management skills.[13] We were told that attitudes toward trainer aircraft are

---

[11] When we presented our research results to AETC, several senior officers expressed the opinion that the *sensor fusion* technology in aircraft like the F/A-22 means that managing information will be easier for the pilot. One of our reviewers (who has flown the F/A-22) agrees, but notes that this means pilots will be expected to fly in more complex, high-threat environments.

[12] Canadians train student pilots from Denmark, the United Kingdom, Singapore, Italy, and Hungary.

[13] Canadian training characteristics are from slides produced by Lieutenant-Colonel Brian Houlgate (2003), Canadian Forces (International Training Programs), and a telephone interview with Mr. Ian Milani (2004) of Bombardier Aerospace Military Aviation Training, who was deeply involved with the development of Canada's training program. Canada bought 22 T-6s (they generate about 20,000 hours per year), and 17 Hawks (which generate 10,000 to 11,000 hours per year). Only about 150 students start pilot training each year, as

similar in the United Kingdom; when the Royal Air Force was seeking a new advanced trainer, its emphasis on cockpit management skills and information management led them to select the Hawk 128 (Milani, 2004).

Air Force Pilot Assessment of Future Missions

Pilots we interviewed from all aircraft communities (fighter, mobility, bomber, and SOF) agreed that future missions will place greater information demands on pilots. There will be more sources of information, more types of sensors, and more ways to communicate with other aircraft and with personnel closer to the battlefield. Almost by definition, situational awareness—the pilot's ability to assess what is going on in relation to the dynamic environment of flight, threats, and mission and the ability to forecast events and decide what to do based on that assessment—will be more difficult, and pilots will need to be able to prioritize the use of information sources and to assess the reliability of that information. Even more operations will be conducted at night, and pilots will have to be comfortable in that environment. Finally, the potential for increased use of UAVs and UCAVs will make force integration more difficult and challenging than dealing with other aircraft with human pilots.

Fighters

For fighter pilots, three capabilities of new aircraft like the F/A-22 will increase the demands on a pilot. Supercruise capability will allow aircraft to fly at supersonic speeds for longer periods. Approaching the battlefield at higher speeds in the supersonic flight regime may complicate information management by decreasing the time available to make decisions. The vectored thrust capability of the F/A-22 will increase the maneuverability of the aircraft (especially in low airspeed, high angle-of-attack regimes) that will, at some point, require new

---

opposed to over 1,000 per year in the United States. The U.S. Air Force will eventually have approximately 454 T-6 aircraft; AETC has about 450 T-38s.

training for pilots. In the cockpit, sensor fusion, in which information from multiple sensors is combined to create a single input to the pilot, may create new demands on the pilot, though some think it will make information management easier.[14]

Mobility Aircraft

Mobility aircraft are flying in higher-threat environments and will continue to do so in the future. Pilots who had flown C-17, C-130, and C-141 aircraft told us that more missions in the future will be flown to unprepared airfields, and they said that flying into and out of such airfields in hostile environments will require greater understanding of the maximum performance characteristics of mobility aircraft. Pilots of future mobility aircraft (or of current aircraft used in the future) will need more knowledge of defensive, and in some cases offensive, countermeasures. Night flying will be more important for mobility aircrews; a phrase we heard at HQ AMC was "night: the new normal environment." Also, if sensor fusion technology is proven in fighter aircraft like the F/A-22, it will likely become the norm in mobility aircraft as well.

Bombers

Bomber aircraft are already flying extended missions—B-2s have flown up to 40-hour missions to Bosnia and Iraq—and these will continue to stress the physical limits of pilots. Beyond this and the increased information-management demands, the only other change in mission cited by those we interviewed was the more complex force integration environment as a result of increased use of UAVs.

---

[14] Unclassified literature indicates that the F/A-22 will be able to sustain approximately nine g's. Both the F-15 and F-16 airframes can withstand nine g's; however, we were told that F-15 pilots generally do not intentionally approach that level for fear of overstressing the aircraft, and F-16 pilots do not spend much time in that regime (the F-16 flight control system also helps prevent over-g's). Because of this, some pilots we interviewed who had not flown the F-22 felt that g-forces might be more stressful in the new aircraft even though the g-limits of the F-22 are not higher than current aircraft. GLOC, which can affect even experienced pilots, is still a danger, and is still a standard briefing item before fighter training missions. Some literature indicates that the F/A-22 anti-g suit will increase a pilot's ability to withstand g forces.

Special Operations

The varied current missions and aircraft of special operations forces make it difficult to generalize about future missions. However, just as the pilots in other communities did, SOF pilots described increases in demands on pilots because of smaller crews in gunships, continued long-range, nighttime helicopter operations, increased communications demands in manned aircraft, and more reliance on unmanned aircraft for surveillance and attack.

Pilots' Assessments of Future Skills

As stated above in the discussion of our research approach, we initially sought to develop a taxonomy of flying skills currently developed in SUPT and to use this as a baseline of skills to which to compare the skills needed for aircraft that will enter the inventory in the near future. The two extremes of such a taxonomy would be a very short list of the most general skills that are currently taught in both the primary and advanced phases of SUPT—such as contact, instrument, formation, and navigation skills—and a complete list of maneuvers that are graded in the SUPT syllabus gradesheets (about 213 items in the T-37 JSUPT syllabus alone).[15] As we continued our interviews, however, it became clear that a list somewhere in between was most useful, because skill descriptions were quite general.

We present two such lists here to show that there has been general agreement over the years about what is important in undergraduate flying training, though there is some difference in the details. The first list, in Table 3.1, is from a 1972 study on the future of undergraduate pilot training from 1975 through 1990. In this table, the skills marked by x's in the columns labeled "Primary" and "Basic" are skills the study group found were needed by all students in undergraduate pilot training. Skills with x's in the "FAIR Basic" column are

[15] This is a count of the items in the T-37 maneuver item files for Basic (32), Contact (48), Instrument (44), Formation (48), and Navigation (41) skills in JSUPT, some of which overlap.

those considered necessary for student pilots who were to be assigned to fighter aircraft (FAIR stands for fighter-attack-intercept-reconnaissance). The last column shows the skills needed for students selected for tanker, transport, and bomber (TTB) aircraft. Skills 21–30 were not taught in undergraduate pilot training at the time the study was conducted.

The second, more general list in Table 3.2 was introduced by Eckerly (1986) in an Air Force Institute of Technology (AFIT) thesis on potential trainer aircraft for students selected for fighter aircraft, and was also used by Chocolaad (2001) in another analysis of requirements for a potential T-38 replacement. While the two lists are different, many of the items in the older list fit into categories in the later one. The major difference between the two is the latter's stress on instruments, flight discipline, and g-awareness.

Our interview approach was to ask general questions about current skills taught in SUPT and skills that would be needed in the future, along the lines of Table 3.2 (see the questionnaire in the Appendix). Questions included the following:

- What do you see as the major changes to Air Force operations in the future compared to the past?
- How will these changes affect the way in which pilots are trained?
- How will these changes affect the type of systems with which current pilots should train today?
- What do you see as some of the most daunting challenges that Air Force pilots will face in the future?
- If you could make recommendations to Air Force leadership with respect to improving the methods and/or systems for pilot training, what would they be?

Table 3.1
1972 Mission Analysis Training Requirements

	Training Requirement	Primary	Basic	FAIR Basic	TTB Basic
1	Ground operations	x	x	x	x
2	Pretakeoff taxi	x	x	x	x
3	Takeoff	x	x	x	x
4	Formation takeoff		x	x	
5	Climb/level off	x	x	x	x
6	Descent approach	x	x	x	x
7	Landing	x	x	x	x
8	Postlanding taxi	x	x	x	x
9	Basic control	x	x	x	x
10	Precision control	x	x	x	x
11	Stall recognition and recovery	x	x	x	x
12	Aerobatics	x	x	x	
13	Unusual attitude recovery	x	x	x	x
14	Pilotage/Dead reckoning	x	x	x	x
15	High/Low-altitude navigation	x	x	x	x
16	Close formation	x	x	x	x
17	Trail formation		x	x	x
18	Communications	x	x	x	x
19	Spin recognition and prevention	x			
20	Emergency procedures	x	x	x	x
21	Tactical formation		x	x	
22	Basic fighter maneuvers			x	
23	Air-to-ground fundamentals			x	
24	Air drop fundamentals				x
25	Radar navigation		x	x	x
26	Crew coordination		x	x	
27	Formation landing		x	x	x
28	Low-level visual navigation	x	x	x	x
29	Collision avoidance	x	x	x	x
30	Decisionmaking	x	x	x	x

SOURCE: Mission Analysis Study Group, 1972.

Table 3.2
Requirements Used in Two Air Force Studies of Trainer Aircraft

Training Requirement	Description
Airmanship	Flexibility Capacity Clearing Flight discipline
Awareness	G awareness Energy awareness Fuel awareness Attitude awareness Orientation awareness
Basic flying (normal and emergency situations)	Takeoffs Landings Climb Descend Turn
Instrument flying	Flying Approaches Landings
Formation flying (multiple aircraft)	Takeoffs Close Rejoins Trail Tactical Basic fighter Approaches Landings
Navigation	High altitude Low altitude
G tolerance	Sustain 6–9 g's Rapid g onset

SOURCE: Eckerly, 1986.

We initially feared that keeping our questions so open ended would make analysis of the responses more difficult. To our surprise, however, interviewee responses on future skills requirements were clear and limited.

All of those we interviewed were of a different mind from the SMi conference participants mentioned above: the first thing Air Force pilots emphasized for the future was that basic skills should always be a priority in undergraduate flying training. By this they usually meant the broad categories of skills shown in Table 3.2, but they stressed the importance of situational awareness. Unless a pilot is comfortable with the basic skills needed for flying the aircraft, he or she will not be able to take on the new information-management demands introduced by modern technology. We frequently heard that flying the aircraft should be second nature, and that undergraduate flying training should stress the development of basic flying skills and situational awareness. In the opinion of those we interviewed, the skills demanded by the more complex aircraft about to enter the inventory include the following:

- the ability to multitask and process information from a variety of sources
- the ability to prioritize what needs to be done
- the ability to recognize when, and when not, to use certain technological tools
- the facility to work through multiple screen options with onboard computer systems (interviewees called this technology layering).

Most of these skills need more emphasis even with today's aircraft. However, pilots we interviewed stressed that most of these skills can be taught in IFF and FTU, that the current SUPT syllabus cannot absorb new skill requirements without either deleting some current skills or lengthening training, that there are few if any skills that they would eliminate from SUPT, and that an SUPT graduate with basic aircraft skills and good situational awareness will be able to learn the new skills in later training. For example, there has been some discussion about providing NVG training at an earlier stage than FTU, where it is currently introduced.[16] Despite the increased importance

---

[16] Helicopter pilots we interviewed said that NVG training is introduced in training at Fort Rucker.

of night operations, most of those we interviewed agreed that SUPT was too early to introduce this skill.[17]

Conclusion

Pilots' assessments of the future Air Force flying environment and mission requirements were consistent with official Air Force planning documents. Their opinions of the changing cockpit environment—specifically, the increased information-management demands introduced by more communications systems and weapons employment capability—are consistent with ideas of the international flying training community. The almost universal opinion of those we interviewed was, however, that the mastery of cockpit-management skills that will be demanded in more advanced aircraft requires basic piloting skills as a foundation, and that there are few, if any, skills that should be added to undergraduate flying training to better prepare student pilots for new aircraft. With the foundation of skills now taught in SUPT, members of the fighter, mobility, bomber, and SOF communities believe that IFF and FTU are better places to introduce the more advanced skills.[18]

---

[17] This is consistent with opinions found in a more detailed study conducted by the Air Force Human Resource Laboratory on the feasibility of introducing NVGs in SUPT and IFF (Martin, 2004).

[18] It is possible that what are considered basic skills can change over time. For example, some of those we interviewed now consider use of GPS a basic skill. The T-6 (2003a) and T-1A (2003f) syllabi include GPS training.

CHAPTER FOUR

Adequacy of the T-38C and the T-1A

We saw in Chapter Two that there is unanimous agreement among those we interviewed that the T-38A/C and the T-1A are adequate for learning the skills needed by current Air Force pilots and that the T-38C is an excellent aircraft for the transition training provided in IFF. The last chapter discussed the nature of future Air Force missions and how they would affect skills needed by pilots from now through 2040. We have also seen that, in the opinion of those we interviewed, the primary skill changes will involve cockpit duties such as multitasking, task prioritization, and information management.

We were somewhat surprised that *virtually no one* in our sample of 230 pilots felt that the nature of these future skills required the consideration of replacements for the T-38C or the T-1A.[1] In their opinion, the T-38C provides adequate training for the undergraduate flying training skills students will need to prepare for future fighter aircraft, and the T-1A provides adequate training for anticipated undergraduate flying training skills needed for future mobility aircraft. This attitude comes from the strong belief that basic flying skills and situational awareness are what matters in undergraduate training, and that IFF and FTU are more appropriate venues for introducing information-management skills that are becoming increasingly important in operational aircraft. If there is no need to replace the T-38C

[1] The only recommendation for a new aircraft came from pilots currently training for the CV-22. Many SOF pilots feel that a tilt-rotor trainer will be necessary for future CV-22 pilots. However, if such an aircraft is not purchased, they felt that going through the helicopter track will be the best preparation.

or the T-1A because they are incapable of providing the skills needed by future pilots, the decision to sustain or replace these two aircraft can be reduced to an economic analysis comparing the costs of sustaining them to the cost of procuring replacement aircraft. Pilots expressed concern about the maintainability of an aircraft as old as the T-38C, but saw no need to replace it if it can be maintained.

Aircraft Modifications

T-38 pilots had several suggestions for improvements to the aircraft; the most important were the following:[2]

- better ejection seat
- antiskid brakes
- better engines in order to have longer sorties and more options for cross-country missions, enable a shorter takeoff roll, and sustain turns and g-forces
- two radios: ultra-high frequency (UHF) and very high frequency (VHF)
- basic autopilot capabilities
- reduced vertical separation minimums (RVSM) capability (TCAS II version 7).[3]

---

[2] These are not in any particular order. The T-38 Weapon Systems Capability Roadmap, however, does show the ejection seat and brake system as high-priority items for funding (Air Education and Training Command/XPPX, 2004b).

[3] "The Traffic Alert and Collision Avoidance System II (TCAS II) is an airborne system that uses active surveillance to alert the pilot of an equipped aircraft to the presence of other nearby aircraft. TCAS II is currently required in the United States on all commercial aircraft with more than 30 seats, and will soon be mandated in many European countries as well. An industry team recently completed the development of requirements for Version 7 of TCAS II. Version 7 is intended to address all known remaining problems and to optimize system performance; the system is also compliant with the [International Civil Aviation Organization] ICAO Standards and Recommended Practices for airborne collision avoidance systems" (Love, 1998).

Some of these suggestions came from pilots who had flown the T-38 in generalized UPT and had not flown the modified T-38C. It is interesting to note, though, that none of them include cockpit changes beyond, or even as advanced as, those that have been made in the T-38C. The ejection seat and brake suggestions are safety related, and are already included as potential improvements in the T-38 Weapons System Capability Roadmap (Air Education and Training Command/XPPX, 2004b), though they are not completely funded. RVSM, implemented in the United States in January 2005, is a plan to reduce the vertical separation of aircraft flying above flight level (FL) 290 from the current 2,000-ft minimum to a 1,000-ft minimum. The purpose is to allow aircraft to safely fly more optimum profiles, gain fuel savings, and increase airspace capacity (Federal Aviation Administration, 2004). Flying under RVSM requires special equipment; the T-38 Weapons System Capability Roadmap indicates that this is prohibitively expensive for the T-38C, and this could in effect limit T-38 flights to altitudes below FL 290 (Air Education and Training Command/XPPX, 2004b). While this will decrease the range available for cross-country flights, it is unclear if this is a significant training problem.

We heard no suggestions in our interviews for modifications to the T-1A. In fact, most people reacted with surprise to the news that the T-1A would need to be SLEPed or replaced by 2020. There is a T-1A Weapons System Capability Roadmap (Air Education and Training Command/XPPX, 2004a), but none of the modification programs in it are funded. The T-1A SPO has recently contracted a supportability and maintainability study for the aircraft; one current concern is the supportability of obsolete parts and the expiration of the Raytheon sustaining engineering support contract in 2009.

Concerns About Post-SUPT Training

Within the context of current SUPT, we have seen that there is no reason to replace the T-1A or the T-38C. However, we heard important concerns about the overall pilot training pipeline and the possibility that graduates of SUPT in the future will be assigned to the F/A-22 and the F-35. The flight regimes currently anticipated for the new fighters include supersonic cruise and significantly improved aircraft maneuvering capabilities under high g forces and at high angles of attack. The F/A-22 System Management Organization (SMO) at Air Combat Command (ACC) indicated that there is more concern about the adequacy of the current IFF and FTU portions of the training pipeline and less concern with the T-38 phase of undergraduate training. The primary cause for this concern is the fact that no two-seat versions of the F/A-22 will be procured by the Air Force. This means that newly trained pilots will eventually be required to learn to maneuver in a sustained nine-g, vectored-thrust environment in solo flight with no instructor pilot (IP) on board to provide essential instruction and safety supervision.[4]

Current Policy with Single-Seat Aircraft

The Air Force currently flies four single-seat fighter aircraft: the F-15C, the F-16C, the F-117A, and the A-10. Both the F-15 and the F-16 have two-seat trainer versions, and training syllabi for both aircraft include several dual sorties in which the student flies with an instructor—not only in transition flights that introduce the student to how the new aircraft handles, but in sorties in which basic fighter maneuvers are introduced, and the student is exposed to higher g-forces than he or she might have experienced in SUPT. F-15, F-16, and F/A-22 pilots all stressed the requirement for students new to fighter aircraft to understand GLOC, and how important it is for an instructor to be present in the event of a GLOC incident.[5] Before

---

[4] The F-35 has no two-seat versions programmed either, and its SMO representative at ACC expressed very similar concerns. However, the F-35 will not become operational for several years after the F/A-22, so we will examine the F/A-22 issues. Presumably they can be thoroughly addressed and fully resolved before new pilots are required to train initially in the F-35.

[5] Experienced F-15 FTU instructors have expressed serious concerns regarding potential dangers associated with new pilots encountering a nine-g environment for the first time in solo flight status because GLOC is not an uncommon occurrence, and many FTU students

students fly solo, they learn about the capabilities of the aircraft with an instructor present in the cockpit.

The F-117A program does not accept immediate graduates of SUPT; applicants must be qualified as leaders of four-ship formations if they come from the F-15 or the F-16. Because F-117A missions are frequently flown single-ship, only experienced decisionmakers are allowed to fly the airplane. Even so, the first six sorties in F-117A training are flown with an instructor pilot flying a chase T-38 in order to monitor how the trainee is doing.

Initial training in the slower A-10 includes sorties in which the trainee is chased by an instructor in another aircraft. We were told that A-10 flight characteristics are relatively easy for a new pilot to adjust to, and there is less concern about the potential for a student to experience GLOC. The g-limits of the A-10 are also lower than for the F-15 and F-16.

Thus, for current single-seat aircraft, two-seat versions exist to help new pilots adjust to new capabilities while with an instructor, only experienced pilots are allowed to fly it, or the aircraft is relatively easy to fly.

As mentioned above, the primary concern of the fighter pilots we interviewed was that it would be unwise for the Air Force to allow SUPT graduates to have their first exposure to nine-g–capable aircraft be in a solo flight because of the possibility of GLOC. When we briefed these results to senior officers at AETC headquarters (Ausink, 2004), we also heard that there were concerns about the fidelity of F/A-22 simulators to aircraft flying characteristics. Because the simulator has not yet achieved the feel of the actual aircraft in certain phases of flight, there are fears that mistakes will be more likely on initial solo flights. For example, scraping the tail during a landing would be extremely expensive, both in terms of repair materials and lost aircraft availability. Many suggested that some type of intermediate training between IFF and the F/A-22 FTU should be developed—perhaps in the F-16—that would allow SUPT graduates more

have regained consciousness on aircraft handling rides to find the aircraft safely under the control of the onboard IP.

time to adjust to high-performance aircraft before flying solo in the F/A-22 or the F-35.[6] This concern is not a problem yet, of course, as only handpicked, experienced fighter pilots are being selected to fly the F/A-22 at this time. However, if the Air Force is to develop an experienced F/A-22 cadre, SUPT students will eventually have to be assigned to that aircraft.

Conclusion

Pilots we interviewed are confident that if the T-1A and the T-38C can be sustained past the year 2020, they will be capable of providing the training needed by student pilots who will be assigned to advanced Air Force aircraft. However, fighter pilots feel that some additional training might be necessary before SUPT graduates can be safely assigned directly to the F/A-22 and the F-35.

---

[6] One of the more creative ideas raised during the September 3, 2004, briefing was to develop software for the F-16 that will allow it to simulate the flight characteristics of the F/A-22, just as the National Aeronautics and Space Administration (NASA) used computer software to allow a Gulfstream to simulate the flight characteristics of the space shuttle.

CHAPTER FIVE

Other Factors Affecting the Replacement Decision

A T-38 Life Study conducted in 2002 by the AETC Studies and Analysis Squadron concluded that if the T-38 were to be replaced between 2020 and 2040, the most cost-effective decision would be to replace it sometime between 2020 and 2025 with a commercial off-the-shelf (COTS) aircraft like the BAE Hawk or with a newly developed aircraft that the study designated the T-XX (Chocolaad, 2001). The assumptions behind this study were first, that the T-38 would be replaced sometime in that 20-year period, and second, that the decision should be based on the weighted consideration of life-cycle cost, reliability, and training effectiveness. A key part of the analysis was the assumption that the COTS aircraft and the T-XX would both be superior to the T-38 in training effectiveness. Interestingly enough, in the sensitivity analysis section of the study, the authors note, "continuing to use the T-38 beyond 2040 would mathematically be the most cost effective option, although it would be practically impossible due to training effectiveness and reliability complications" (Chocolaad, 2001, p. 10).

What we have learned in our interviews is that training effectiveness limitations of the T-38 are overstated and, in the context of current SUPT, the replacement decisions for the T-38C (and the T-1A) can be made on cost comparisons alone.[1] A detailed analysis comparing the costs of SLEPing the T-38C to the purchase or

[1] This is not a criticism of the T-38 Life Study; it simply appears that some of its assumptions may no longer be valid.

development of a new trainer might show that sustaining the T-38C is the cheaper option in the context of SUPT. Based on current undergraduate flying training demands, we simply did not find a compelling need for replacement aircraft, and any such decision could therefore be based upon economic considerations. This particular conclusion relies upon several important assumptions and must be interpreted within its very narrow context. Thus, it should be regarded as quite distinct from the general conclusion that AETC will require no replacement aircraft by the year 2025. Economic issues associated with aging aircraft are quite complex, and there are also several potentially compelling operational issues that could have a significant effect on the analysis of economic factors.

Economic Issues Associated with Aging Aircraft

It is important to recognize at the outset that the T-38 could become one of the oldest airframes ever flown in mainstream training operations. The current T-38 fleet averages almost 14,000 flying hours per airframe, which is twice the original design service life (DSL) for the trainer. Moreover, if no replacement aircraft is programmed and the T-38 is operated until as late as 2040, the Air Force could be training a sizable portion of its new pilots in airframes that are almost 80 years old. Although the T-1A is of more recent origin, it still raises similar issues if no replacement is programmed.[2]

Recent RAND research supporting the Project AIR FORCE Aging Aircraft Project documents some of the complexities associated with analyzing the economic issues involved in the replacement of aging aircraft. The general conclusion is that "...the Air Force should repair, rather than replace, an aging system if and only if the

---

[2] T-38 flying hour information is from the Air Force's Aging Aircraft Technologies Team (AATT) survey data published in its 2002 report (U.S. Air Force, 2002). Conversations with the T-1A SPO indicate that there are DSL concerns with that aircraft as well because the increased numbers of practice landings and other military operating issues are more demanding than the civil flying profiles assumed in the manufacturer's original estimates. The AETC staff provided additional data in this paragraph.

availability-adjusted marginal cost of the existing aircraft is less than the replacement's average cost per available year."[3] This research provides a mathematical model to calculate the decision process required and further enumerates the parameters that must be estimated in order to use the model. The researchers then urge caution in interpreting their model results because of potential difficulties in obtaining accurate parameter estimates and acknowledge that the decision process is often clarified when there is a significant increase in operational capability associated with the replacement aircraft (Keating and Dixon, 2003).

Companion research documents show how the age of the aircraft fleet relates to maintenance and modification workloads as well as material consumption. The research also confirms that these factors typically exhibit late-life growth as the aircraft ages, but the growth rate itself depends upon a number of additional parameters, including flyaway costs and the specific workload category. These results are incorporated into additional models that are designed to allow the Air Force to better estimate the safety, aircraft availability, and cost implications associated with retaining aircraft fleets for heretofore-unprecedented time periods (Pyles, 2003).

The economic analysis is further complicated by several operational issues that help define the undergraduate flying training demands emphasized above. For example, the improved training capability of a new aircraft may be difficult to evaluate objectively, and there is no *a priori* evidence that the basic flying skills that student pilots need to develop would exhibit dramatic improvement in a replacement aircraft (over the existing options). The improved operational capabilities expected from the F/A-22 and the F-35 might require modifications in the existing training pipeline, to ensure a seamless transition into the new systems, but our study indicated that potential modifications would be more appropriate in training phases that currently are subsequent to undergraduate training. These potential modifications to the training pipeline could still become

---

[3] The quote is taken from the document abstract.

dominant in a replacement aircraft analysis, because a replacement airframe might be used in both undergraduate flying training and postundergraduate training. Other factors that may be germane in the replacement analysis involve future Air Force policy options and the flexibility associated with future pilot production requirements.

New operating regimes are currently associated with F/A-22 and F-35 fighter replacement aircraft rather than nonfighter aircraft, so the T-38 training track might be most affected by new training needs, but this does not preclude future technological developments in mobility aircraft as well.

The Tracking Decision

Pilots in the fighter, bomber, and mobility aircraft communities all thought that the tracking decision in SUPT is made too early. Some felt that students who had completed T-37 or T-6 training did not have enough experience to make informed career decisions about what they wanted to fly. Others felt that performance in the primary phase of training was insufficient for instructors to make the best decision about what track was best for a student.[4] There are at least three approaches to addressing this issue, and all of them would affect the replacement decision for the T-38C.

First, the Air Force could extend primary phase training in the T-6. This could allow both students and instructors to make better decisions about later tracking, especially if new sorties introduced more formation or fighter-type maneuvers that would give a taste of the type of flying that would be done in the advanced phase in the

[4] There are many different opinions about the best time to track students. Ideally, one would be able to track them before they have even flown an aircraft, so that they are only trained in the type of aircraft they will fly operationally. RAF pilots told us that the United Kingdom is moving the tracking point earlier and earlier in training, so that students know what type of aircraft they will be assigned to after just 50 hours of flying in a program analogous to IFT in the United States. The Israeli Air Force apparently tracks students to fighter or mobility aircraft after only 25 sorties in two different aircraft, although the screening process there before a student flies at all is much more involved than in the United States (Hays, 2002).

fighter track. If, in order to retain the current length of SUPT, the increased hours in the T-6 resulted in fewer hours in the T-1A and the T-38C, the demands on those two airframes would decrease, and their lifetimes would be extended. This would probably weight the decision toward retaining the two aircraft.

A second approach to the tracking problem, and one that was mentioned several times in our interviews, would be to give all student pilots some experience in the T-38C. Many felt that this would help weed out students who would immediately recognize that they could not handle a faster aircraft, and would also inspire students who thought they wanted to fly mobility aircraft to become interested in the fighter track. This approach would much more than double the number of students who spend at least some time in the T-38, which would affect long-term airframe demands and most likely decrease the probability that the T-38 could last through 2040.[5]

A third approach would be to return to generalized UPT so that all students fly the T-6 and the T-38C (or a follow-on replacement). Only a few pilots we interviewed suggested such an approach, but not all of them were senior officers who had themselves graduated from that program; several SUPT graduates thought there would be benefits to a single-track program. Obviously, such a change would put huge demands on the T-38, and it is likely that the available T-38 airframes would be unable to meet the task. Taking this route would almost certainly require the purchase of a replacement for the T-38C, but the cost of doing so could be affected by some changes in FTU discussed below.

---

[5] According to the T-1A Weapons Systems Capability Roadmap (Air Education and Training Command/XPPX, 2004a), when pilot production is 1,100 students per year, approximately 650 students will train in the T-1A. If all students had some time in the T-38, the initial student load in the aircraft would increase from 450 to 1,100. There might not be enough T-38s to meet the programmed flying training (PFT) requirements in this case.

Flexibility to Meet Future Air Force Training Needs

Potential changes to undergraduate flying training will be based not only on the need to teach new skills, but also on the total demand for pilots. The ability to satisfy changing pilot production requirements could influence the decision to modify undergraduate training or to obtain new aircraft.

The Air Force experienced problems meeting pilot production goals throughout the post-Cold War period because of decisions made about how to draw down military manpower. The resulting pilot shortfalls in the lower ranks now threaten to limit the Air Force's ability to develop sufficient numbers of officers with the operational knowledge and mission experience to provide essential future guidance and leadership.[6] Some imbalances may have been aggravated by changes in undergraduate flying training: graduates of generalized UPT who had initially flown nonfighter aircraft were allowed to apply for empty fighter slots, but SUPT graduates who had taken the nonfighter track were not, so flexibility in changing aircraft was limited. In addition, problems resulted from a continuing requirement following the drawdown to operate training pipelines at their maximum production capacity in order to rebuild, with no buffer available to mitigate unforeseen issues.

If the T-38 and the T-1A are not replaced, normal aircraft attrition during flying operations will mean that pipeline capacities will continue to decrease. Developing or purchasing replacement aircraft would eliminate this problem, but the economic consequences of buying additional capacity or flexibility in this manner must be examined thoroughly.

---

[6] RAND analysis that discusses the pilot training and inventory issues generated by these postdrawdown decisions is documented in Taylor, Moore, and Roll (2000) and Taylor et al. (2003). This continues to be an area of active research.

Potential FTU Changes Because of the F/A-22 and F-35

We noted in the last chapter that many current fighter pilots are concerned about the dangers of SUPT graduates being assigned to the F/A-22 and being exposed to an aircraft capable of nine g's for the first time in a solo flight. We also heard concerns about new graduates flying a very expensive and sophisticated aircraft and, through inexperience, making errors that could cause minor, but expensive, damage. As mentioned in the last chapter, some of those we interviewed suggested that SUPT graduates assigned to the F/A-22 should go through some type of program in the F-16 so they could be exposed to some of the features of the F/A-22 (high-g capability, fly-by-wire with side control stick) while flying with an instructor. Cost factors that would require analysis if this sort of modification were to become an essential part of the training pipeline would be considerable, e.g., the added costs of delaying the ongoing F-16 divestiture as well as the obvious costs associated with the service life extension and increased sustainment costs that would be necessary over and above the normal operating costs. The latter would need to include the training costs associated with maintaining and operating a separate aircraft fleet to conduct a very limited amount of training, although the additional aircraft transition (into the F-16) required for the IFF students would be likely to extend the number of flying hours required per pilot well beyond the current 17 hours. This could generate additional increases in aggregate flying hour costs. The aircraft replacement analysis is further complicated by the fact that the T-50, a trainer based on the F-16 and designed for undergraduate flying training, is already in its full-scale development phase, and may have the potential to generate sizable economies of scale.

The T-50 is a joint venture between Lockheed Martin and Korea Aerospace Industries (KAI) that is expected to go into production in the near future (SPG Media PLC, 2004). The two companies announced in July 2004 at the Farnborough International Air Show that the full-scale development program is "...on track for completion at the end of 2005." They also indicated that their aircraft production schedule should support initial Korean training operations in 2006 and should continue for at least 25 years. We discuss this aircraft here because it is the most advanced jet trainer currently in use or under full-scale development, and it would be very likely to have the capability to replicate any IFF training that could be conducted in the F-16 itself. Thus, if it were to be used in undergraduate flying training as well as IFF, it could conceivably eliminate the requirement for a distinct IFF aircraft (and presumably, therefore, the associated costs outlined in the preceding paragraph). A thorough cost analysis, however, would require its detailed comparison to other available aircraft and options, including the T-38 (and/or two-seat versions of the F/A-22 and F-35), for the required training phases.[7]

This approach would require cost analysis of the entire pilot training pipeline: if a new trainer aircraft is the best way to prepare SUPT graduates for the F/A-22, it might make sense to use the aircraft in IFF as well, and if replacing the T-38C in IFF is a good idea, it might be worth considering the possibility of replacing the T-38C in SUPT. Finally, if a new aircraft is introduced in SUPT, the airframe limitations of the T-38 are no longer an issue, and it might be reasonable to consider returning to a generalized approach to undergraduate pilot training. A larger purchase of aircraft would make them more affordable.

Unmanned Aerial Vehicles

As we saw in Chapter One, there will be a significant increase in the number of UAVs in the inventory by 2025. Some defense analysts think that one-third of the Department of Defense's (DoD's) aircraft fleet will consist of UAVs by 2010 (Pae, 2004). How these UAVs are deployed in the Air Force CONOPs (concept of operations) may

---

[7] The bulk of the information in this paragraph is taken from the KAI Web site, www.koreaaero.com/english, and the quoted text is from a news reprint found there (Korea Aerospace Industries, 2004). KAI was established in 1999 with the consolidation of Samsung Aerospace, Daewoo Heavy Industry, and Hyundai Space and Aircraft.

affect pilot training. Most obviously, if the UAV inventory increases at the expense of other types of aircraft, the number of pilots needed annually may decrease. Lower pilot production will decrease the demands on current airframes and make it easier to sustain the T-1A and the T-38C beyond 2020.

CONOPs are important because UAVs do not necessarily need pilots to monitor them. For example, the Global Hawk, which provides intelligence and reconnaissance imagery, can be programmed to taxi, take off, fly to its reconnaissance location, return to base, and land without human intervention. A large fleet of similar UAVs could significantly decrease the number of pilots needed in the Air Force (U.S. Air Force, 2003a). On the other hand, the Predator UAV, which performs interdiction and armed reconnaissance missions, is flown by a pilot stationed in a ground-control station (U.S. Air Force, 2001). If more UAVs in the future require pilots to fly them in real time, the pilot training for these specialists might have to be different from SUPT.

Predator pilots we interviewed told us that the aircraft has three missions: killer scout, which assists fighter pilots in determining targets; intelligence, surveillance, and reconnaissance (ISR); and surface air attack (SAT). A flying background can be useful for a person to be an effective Predator operator: people with backgrounds in heavy aircraft apparently find the takeoff and landing phases easier than people from other backgrounds do. A fighter background and familiarity with weapons employment is useful for the aircraft's mission. Nonetheless, we were told that being an Air Force pilot or having a commercial license is probably not necessary to fly the Predator.[8] At most, going through a primary phase in the T-6 followed by an FTU for the UAV would be sufficient. One person suggested that a completely new UAV Air Force Specialty Code for enlisted personnel and a new training system that does not include flying would be sufficient. In this person's opinion, officers in UAV squadrons should still be pilots, however.

---

[8] However, we also heard that the Federal Aviation Administration (FAA) may require operators of UAVs in U.S.-controlled airspace to have a pilot's license.

Simulators

The 1972 Mission Analysis Group study on future pilot training concluded that "substantial reductions in flying time can be made by using simulation and that the pilot quality can be increased as a result" (Mission Analysis Study Group, 1972, p. 57). Attitudes in our interviews toward simulators were mixed.

There was universal praise for the simulators used in T-6 and T-38C training, and students in fighter FTUs also felt that simulators used in F-15 and F-16 training provided excellent preparation for flying the aircraft. Pilots at the F-16 FTU told us that a robust simulator program makes the transition to the aircraft easier, but that simulator missions could not replace actual aircraft sorties. Pulling g's, the fear of death from making a mistake, heat, radio traffic, and equipment failures were all described as difficult or impossible to simulate. These were listed as reasons not to replace aircraft sorties with simulator sorties for trainees in the F-16. On the other hand, the distributed mission operations (DMO) program, which allows pilots in simulators at geographically separated locations to perform missions together, was thought to be an excellent way for operational pilots to hone their skills.[9]

Mobility pilots had similar comments about the lack of a fear factor in simulators, but seemed more inclined to accept some substitution of simulator time for aircraft time. We were told at HQ AMC that the simulator for the new C-130J will be almost as good as commercial simulators, and will allow three aircraft sorties to be replaced by simulator sorties.[10] The C-17 FTU syllabus includes only

---

[9] A briefing by Wright-Patterson AFB's Aeronautical Systems Center Training Systems Product Group (U.S. Air Force, 2004b) describes the Mission Training Centers (MTCs) used in DMO as high-fidelity, locally networked simulators (two- to four-ship) that are linked in a wide area network to simulate large exercises. This creates a virtual battlespace for people to rehearse missions with many aircraft.

[10] The FAA defines four levels of simulators, A through D. Level D simulators require (among other things) night, dusk, and day visual displays, six-axis motion (three axes of tilt plus three axes of motion), ability to simulate the look and feel of runway contaminants, realistic cockpit noise, and operating radar (Federal Aviation Administration, 2000). According to mobility pilots we interviewed, the FAA accepts some types of simulator

three aircraft sorties, with 46 missions in procedural trainers and simulators. Mobility pilots affirmed the value of DMO training.

Some pilots we interviewed speculated that improved simulator technology might make it possible for SUPT graduates to safely begin training in the F/A-22, but we also heard that the F/A-22 simulator had not yet achieved the fidelity necessary to give new pilots the feel of the actual aircraft. Although one person we interviewed said that the F/A-22 is "easier to fly than a Cessna," we still heard concerns about landing the aircraft and managing g-forces on the first flight without an instructor.

Overall, we conclude that pilots of nonfighter aircraft are more comfortable with the idea of simulator sorties replacing aircraft sorties than fighter pilots are. All agree that modern simulators are important aids to initial training and for maintaining skills (in DMO exercises, for example).[11] It does not appear, though, that increased use of simulators in SUPT will lead to much reduction in actual aircraft time.[12]

Conclusion

We can summarize the factors discussed in this chapter that affect the replacement decision by placing them into four categories:

- Strategy: Increased use of UAVs, and other decisions that affect the number of manned aircraft needed to perform Air

training as equivalent to aircraft training, and the first time some commercial pilots fly an actual airliner is as a copilot on a flight with passengers.

[11] In many discussions with pilots, co-author William W. Taylor has found that many pilots think simulators are better for improving the skills of operational pilots than for developing skills in new pilots.

[12] On the other hand, the T-6 syllabus has 12 more simulator hours in the instrument phase of training and three more hours of instrument training in the aircraft than the T-37 does. Some T-6 instructors indicated that the extra simulator time made the extra T-6 aircraft hours redundant. However, if the extra instrument hours in the aircraft were eliminated, they would want them moved to another phase of training rather than eliminating them altogether and shortening the overall length of the T-6 course of instruction.

Force missions, could affect the demand for pilots as well as the nature of pilot training.

- Policy: A desire for increased flexibility in the assignment of Air Force pilots could influence what skills need to be taught in pilot training and the demand on training airframes.
- Training: More advanced simulators and other improved approaches to ground-based training could affect the number of flying training hours required and the demands on training airframes. Any changes in the timing of the tracking decision will also affect the demands on the T-38. Changes in the requirements for FTU could also affect the decision to retain or replace current training aircraft.
- Budget: Understanding the economics of aging aircraft is important, since it is difficult to predict the costs associated with maintaining older airframes. In addition, any decision about a follow-on trainer aircraft must take into account the costs associated with the classroom instruction, computer-based training, simulators, and other ground-based training necessary to augment what is taught in the aircraft.

CHAPTER SIX

Next Steps

We have concluded that in the context of current SUPT, the T-38C and the T-1A are capable of developing the skills that Air Force pilots will need in order to fly the aircraft projected for the Air Force inventory through 2040. As a result, the decision to replace them can be based solely on economic considerations; it is difficult to argue that replacement aircraft are necessary because current aircraft lack the requisite training effectiveness. We stress that this conclusion is limited to the context of current SUPT. All of the factors mentioned in Chapter Five can affect the entire pilot training pipeline, changes in which would affect the decision to sustain or replace the T-1A and the T-38C.

Any decision must thus also consider the costs and effects of changes on the entire training pipeline. For example, it would not make sense to make changes in undergraduate training that increase the pilot production rate if the FTUs are not capable of handling more students as well. Beyond the training pipeline, an attempt should be made to consider the potential effects of poor screening or tracking on long-term retention of pilots. In the Hoss study on A-10 FEBs mentioned in Chapter Three (Hoss, 2003), one student was in the top third of his T-37 class, but was in the bottom half of his T-38 class and ended up failing training in the A-10. The pilot was given a waiver to continue as a pilot in the C-130. This may or may not have an effect on this person's motivation to stay in the Air Force, but this is certainly a case in which better tracking to a nonfighter aircraft in the first place could have saved resources.

AETC should consider using the following approach to finalizing its replacement decision.

1. As a baseline, determine the cost of continuing SUPT and IFF in their current forms by SLEPing the T-38C and the T-1A. At the same time, determine the cost of retaining trainer versions of the F-16 in order to use them in a pre–F/A-22 FTU program that will expose new pilots to high sustained g-forces in the presence of an instructor before they fly solo in the F/A-22. It is likely that only a small number of F-16s would be necessary for such training, but infrastructure costs per airframe increase as the number of airframes decreases. These costs together provide a baseline cost of a training pipeline with no new aircraft. As part of this analysis, the possibility of modifying F-16s to simulate F/A-22 flying characteristics should also be considered.
2. Compare this to the cost of continuing SUPT and IFF in their current forms but with replacement aircraft for the T-1A and the T-38C. For T-38C replacement aircraft, it makes sense to consider some version of the BAE Hawk or the T-50. This comparison should also include the cost of using the replacement for the T-38C in IFF and in a pre–F/A-22 FTU program.
3. As a first excursion from current SUPT, the possibility of extending T-6 training before the tracking decision is made should be considered. This would decrease the demands on both the T-38C and the T-1A in SUPT, which could mean they would last longer even without a SLEP.
4. As a second excursion from current SUPT, the effect of allowing all students to fly the T-38C (in order to expose them to a higher-performance aircraft before tracking) should be analyzed.[1]
5. Finally, the costs of returning to single-track UPT should be considered, first in a version using a SLEPed T-38C,[2] and second

[1] Lt Gen Baker, Vice Commander of AMC, and others suggested that all students should have as many as eight sorties in the T-38C before the tracking decision is made.

[2] Chocolaad (2001) considered this case as part of his study, and determined that the Air Force would have to purchase T-38s from other U.S. (and perhaps foreign) organizations in

with a replacement for the T-38C. In the second case, the replacement aircraft would be used for both the advanced training phase and IFF, and possibly for a pre–F/A-22 version of FTU. The single-track option would also introduce a tanker/transport version of IFF: after single-track UPT, graduates could attend a short course in the T-1A before going on to FTU (many people we interviewed think that current T-1 training is too long). This option will be interesting to consider, since it would replace only one aircraft (instead of replacing or SLEPing two), and might provide an option for interim training for new pilots who have been assigned to the F/A-22. Thus, while there is no compelling training reason to return to single-track UPT, cost considerations might make doing so an attractive option.

In all of the cost studies suggested above, the potential reduction in the need for pilots because of the introduction of more UAVs must be considered. An important concern is that the Air Force may not know for some time whether specific changes to the current programs will be required to prepare new pilots to transition to the F/A-22 because only experienced fighter pilots are currently programmed to make the transition. It is important to note, however, that only a limited period of time is available to achieve the essential budgetary programming that would be required to buy a replacement for the T-38. Excessive delays will probably make the replacement option more difficult to incorporate into this programming process.

While the skills required for pilots in the future do not mandate the replacement of the T-38C and the T-1A, the structure of the pilot training pipeline and the number of pilots produced will have a large effect on the decision to sustain or replace them.

---

order to meet the expanded sortie requirements. Because of attrition, he concluded that the T-38 would not be able to meet sortie requirements after 2020. The study took into account the costs of some upgrades to the T-38 (including those for the T-38C), but it does consider a more extensive SLEP that might make the aircraft more reliable or sustainable.

APPENDIX

Interview Questionnaire

For Senior Leaders and Planners

1. What changes are planned that will affect the skill requirements of Air Force pilots in the 2020–2030 timeframe? (Consider all pilots, e.g., fighter, bomber, tanker/transport, UAV, test, surveillance.)
   a. Range of missions (air-to-air combat, air-to-ground, escort, decoy, bombing, gunship, refueling, surveillance, and electronic combat)
   b. A/C employment (single aircraft, aircraft in formation, aircraft with deployable UAVs onboard or in formation with UAVs)
   c. Physiological demands (g forces, sleep deprivation)
   d. Mental demands (information management, weapons employment, multitasking, mission and deployment length, communication skills, formation and strike force management)
   e. Other
2. How do you expect aircraft and the aircraft environment in the 2020–2030 timeframe to differ from what they are now?
   a. New technology (e.g., F/A-22 vectored thrust)
   b. Increased technology (HUD, GPS and other navigation systems, satellite and EC-135 communications systems, electronic warfare systems, and advanced weapons employment (air-to-air and air-to-ground)
   c. Other

For Pilots in the Field or at Formal Training Units (FTUs)

Aircraft Skills

Current Training

1. What are the major differences between flying the aircraft used in training (SUPT, IFF, or Altus) and flying the current operational aircraft (e.g., differences in landing difficulty between the T-38 and the F-15/F-16/A-10)?
2. What skill deficiencies (if any) do you observe in new pilots that come to you from SUPT? IFF (for fighters)? Altus (for heavies)? How can these deficiencies be eliminated?
3. What advanced motor skills are taught in FTU that are not taught in IFF or SUPT?
4. What advanced cockpit skills (instrument reading, interpretation and action, button-pushing skills) are taught in FTU that are not taught in SUPT, IFF (fighters), or Altus (heavies)?
5. What advanced cockpit crew skills are taught in FTU that are not taught in IFF, SUPT, or Altus? (Note that crew skills for fighter pilots could include communication with other aircraft in formation.)
6. Are there any skills developed in the aircraft used in training (SUPT, IFF, or Altus) that are not needed in the operational aircraft to which a trainee has been assigned? If so:
   a. Which ones?
   b. Do you think learning these skills slows down training to a significant degree? If so, please elaborate.
   c. Do such unnecessary skills prepare pilots for other types of skills or challenges they will face after training?
   d. Are there other reasons we might want to retain them (for example, to test trainees' commitment, endurance, or adaptability to changes in aircraft)?
7. Which skills are best learned by flying the operational aircraft to which the trainee is assigned? Is there any way the training aircraft could be modified to teach any of these skills adequately? Is there any way to use simulators to teach any of these adequately?

8. What skills are best learned in a simulator? Is there any way simulators can be used to improve the efficiency of training?
9. Do you observe (or anticipate) any differences in the ability to adjust to new aircraft based on pilot age or experience?
10. Based on your observations of new pilots that come to your training, are there any aircraft or cockpit skills that you think should be stressed more in IFF? In SUPT? In Altus?

Future Training

1. What new motor skills (hands-on flying skills) do you think will likely be required in future aircraft that are not currently taught in FTU, IFF, or SUPT?
2. What cockpit skills (instrument reading, interpretation and action; button-pushing skills; or connectivity [internet-type connections in the cockpit]) do you think will likely be required in future aircraft that are not currently taught in FTU, IFF, or SUPT?
3. What cockpit crew skills (crew coordination, emergency response) do you think will likely be required in future aircraft that are not currently taught in FTU, IFF, or SUPT?

Environment

What changes would you like to see in future aircraft or the aircraft environment that would be fundamentally different from what it is now?

1. F/A-22 has vectored thrust. What else might be developed? *[If suggestions are offered, ask the following:*
   - *What benefit would that provide?*
   - *How great would the impact be on your ability to carry out your missions?]*
2. In Vietnam, pilots had to handle limited weapon systems and radar; now they have HUD, GPS, and other navigation systems, satellite and EC-135 communications systems, electronic warfare

systems, and advanced weapons employment (air-to-air and air-to-ground). What new demands might come into play?

For Introduction to Fighter Fundamentals (IFF)

Aircraft Skills

Current Training

1. What skill deficiencies (if any) do you observe in new pilots that come to you from SUPT? How can these deficiencies be eliminated?
2. Are there any aircraft or cockpit skills that you think should be stressed more in SUPT?
3. What new motor skills can be taught in the T-38C that could not be taught in other trainers?
4. What new cockpit skills can be taught in the T-38C that could not be taught in other trainers?
5. What advanced motor skills are taught in IFF that are not taught in SUPT?
6. What advanced cockpit skills (instrument reading, interpretation and action, button-pushing skills) are taught in IFF that are not taught in SUPT?
7. What advanced cockpit crew skills are taught in IFF that are not taught in SUPT? (Note that crew skills for fighter pilots could include communication with other aircraft in formation.)
8. Are there any aircraft, cockpit, or crew coordination skills that you think should be stressed more in SUPT (including those that might require a new aircraft in SUPT)?
9. Do you think the current approach to tracking students in SUPT helps or hinders readiness for IFF?
10. Do you observe (or anticipate) any differences in the ability to adjust to new aircraft based on pilot age or experience?

Future Training

1. What motor skills (hands-on flying skills) do you think will likely be required in future aircraft? Are there any that are not currently taught in FTU? In SUPT?
2. What cockpit skills (instrument reading, interpretation, and action; button-pushing skills; or connectivity [internet-type connections in the cockpit]) do you think will likely be required in future aircraft? Are there any that are not currently taught in FTU? In SUPT?
3. What cockpit crew skills (crew coordination, emergency response) do you think will likely be required in future aircraft? Are there any that are not currently taught in FTU? In SUPT?

Environment

What changes would you like to see in future aircraft or the aircraft environment that would be fundamentally different from what it is now?

1. F/A-22 has vectored thrust. What else might be developed? *[If suggestions are offered, ask the following:*
   - *What benefit would that provide?*
   - *How great would the impact be on your ability to carry out your missions?]*
2. In Vietnam, pilots had to handle limited weapon systems and radar; now they have HUD, GPS, and other navigation systems, satellite and EC-135 communications systems, electronic warfare systems, and advanced weapons employment (air-to-air and air-to-ground). What new demands might come into play?

Bibliography

AETC. See Air Education and Training Command.

Ahearn, Dave, "New Cost Estimates Might Bolster Raptor Program," *Defense Today*, July 26, 2004.

Air Education and Training Command, "T-38C Programs," undated briefing.

———, "Major Changes in Undergraduate Pilot Training 1939–2002," 2002a. Online at http://www.aetc.randolph.af.mil/ho/upt_changes/upt_preface.htm as of October 2004.

———, "T-38A Specialized Undergraduate Pilot Training," AETC Syllabus P-V4A-A (T-38A), 2002b.

———, "USAF F-15 Initial Qualification Course," AETC Syllabus F15ACB00AT, 2002c.

———, "T-6 Joint Primary Pilot Training (TIMS), "AETC/CNATRA Syllabus P-V4A-J (TIMS), 2003a, with Change 1.

———, "USAF F-16C/D Initial Qualification Course (Luke)," AETC Syllabus F16C0B00PL, 2003b.

———, "ENJJPT Fact Sheet," Randolph AFB, Tex., 2003c.

———, "T-37 Specialized Undergraduate Pilot Training," AETC Syllabus P-V4A-A (T-37), 2003e.

———, "T-1 Joint Specialized Undergraduate Pilot Training/Fixed-Wing Transition," AETC Syllabus P-V4A-G/F-V5A-Q, 2003f, with Change 1.

Air Education and Training Command/XPPX, "T-1A Weapons System Capability Roadmap," DRAFT Revision 1, Randolph AFB, Tex., 2004a.

———, "T-38 Weapons System Capability Roadmap," DRAFT Revision 1, Randolph AFB, Tex.,. 2004b.

Air Training Command, "Comparison of UPT Generalized vs Specialized," HQ ATC DCS Operations, March 5, 1976.

Ausink, John A., "Flying Training 2020: Assessing the Impact of Future Operations on Trainer Aircraft Requirements," briefing to General Donald G. Cook, September 3, 2004.

The Boeing Company, "Boeing Avionics Help Guide F-22 Missile to Its Target," press release, Seattle, September 24, 2001. Online at http://boeing.com/news/releases/2001/q3/nr_010924n.htm as of March 16, 2005.

Cahlink, George, "War of Machines," *GovExec.com*, July 15, 2004. Online at http://www.govexec.com/features/0704-15/0704-15s5.htm as of March 16, 2005.

Chocolaad, Christopher A., *T-38 Life Study*, AETC Studies and Analysis Squadron, Randolph AFB, Tex., October 24, 2001.

Davis, Paul K., "Strategic Planning Amidst Massive Uncertainty in Complex Adaptive Systems: The Case of Defense Planning," *InterJournal,* Complex Systems, No. 375, 2002.

DCS/Operations HQ Air Training Command, *Comparison of UPT: Generalized vs Specialized*, Randolph AFB, Tex., 1976.

Deale, Thomas, *On the Fields of Friendly Strike .. . The Dichotomy of Air Force Doctrine and Training Involving Real-Time Targeting*, Air University, Maxwell AFB, Ala., 1999.

Deptula, Maj Gen Dave, "Turning Vision into Reality: Air Force Transformation," June 27, 2001. Online at http://www.dfi-intl.com/Shared/AirSpaceSeminars/General_Deptula_transformation_Brief__27_June_01.pdf as of March 15, 2005.

Eckerly, John M., Roland M. Sasscer, Theodore V. Shropshire, Brian G. Woika, Raymond H. Young, *A Systems Engineering Method for the Design and Selection of a Reconnaissance-Attack-Fighter Pilot Training System, Volume II, Final Report*, AFIT/GSE-86D, Air Force Institute of Technology, Wright-Patterson AFB, Ohio, 1986.

Emmons, Richard H., *Specialized Undergraduate Pilot Training and the Tanker-Transport Training System*, Air Training Command History and

Research Office, HQ Air Training Command, Randolph AFB, TX, July 1991.

Endsley, Mica R., "Measurement of Situation Awareness in Dynamic Systems," *Human Factors*, Vol. 37, No. 1, 1995, pp. 65–84.

Ennett, Jimmy, *The Impact of Emerging Technologies on Future Air Capabilities*, DSTO-GD-0186, Defence Science and Technology Organisation, Melbourne, Australia, 1999. Online at http://www.dsto.defence.gov.au/corporate/reports/DSTO-GD-0186.pdf as of March 15, 2005.

Fazzini, Paul, "New Manual Gives Mobility Crews 'Go-to-War' Guidance," *Air Force Link*, Scott AFB, Ill., August 19, 2004.

Federal Aviation Administration, "General Requirements for Simulators," May 25, 2000. Online at http://www.faa.gov/nsp/documents/Sim_Levels.doc as of March 16, 2005.

———, "Reduced Vertical Separation Minimum." Online at http://www.faa.gov/ats/ato/rvsm1.htm as of September 22, 2004.

Fraser, Maj Gen William M. III, "Training for Today and Tomorrow, briefing, March 11, 2004.

Grant, Rebecca, "The Redefinition of Strategic Airpower," *Air Force Magazine*, Vol. 86, No. 10, 2003, pp. 33–38.

Hallion, Richard P., "Precision Guided Munitions: The New Era of Warfare," *Air Power Studies Center Paper Number 53*, Australia, 1996.

Hays, Michael D., *The Training of Military Pilots: Men, Machines, and Methods*, School of Advanced Airpower Studies Thesis, Air University, Maxwell AFB, Ala., 2002.

Hoss, 1st Lt Kim, "T-38 Predictors Study," briefing, AETC Studies & Analysis Squadron, November 22, 2002.

———, "A-10 FEB Review," briefing, AETC Studies & Analysis Squadron, August 27, 2003.

Houlgate, Lt Col Brian, *JSF and Pilot Training in Canada*, briefing, Lockheed Martin Information Systems, Orlando Florida, June 19, 2003.

Jefferson, Senior Airman Oshawn, "Boeing Unveils UCAV," *Air Force Print News*, September 28, 2000. Online at http://www.fas.org/man/dod-101/sys/ac/docs/man-ac-ucav-000928.htm as of March 28, 2005.

Joseph, Capt Steven G., *Alternatives for Future Undergraduate Pilot Training,* ATC Operations Analysis Report 77-1, 1977a.

———, Specialized UPT Screening, Flexibility, and Other Issues, ATC Operations Analysis Report 77-4, 1977b.

Jumper, Gen John P., Air Force Chief of Staff, Remarks at the Air Force Association Air Warfare Symposium, Orlando, Fla., February 13, 2003.

"Jumper: Air Force Will Determine JSF Inventory Needs in 2006 POM," *Inside the Air Force,* July 30, 2004, p. 1.

Keating, Edward G., and Matthew C. Dixon, *Investigating Optimal Replacement of Aging Air Force Systems,* Santa Monica, Calif.: RAND Corporation, MR-1763-AF, 2003.

Khalilzad, Zalmay, and Jeremy Shapiro, eds., *Strategic Appraisal: United States Air and Space Power in the 21st Century*, Santa Monica, Calif.: RAND Corporation, MR-1314-AF, 2002.

Korea Aerospace Industries, "LM, KAI Say T-50 Supersonic Trainer Developing Rapidly," *KAI News*, August 10, 2004. Online at http://www.koreaaero.com/english/cyberpr/sub1/sub1_view.php?pidx=37&page=1 as of March 16, 2005.

Lambeth, Benjamin, "Technology and Air War," *Air Force Magazine*, Vol. 79, No. 11, 1996. Online at http://www.afa.org/magazine/nov1996/1196techairwar.asp as of March 15, 2005.

Lambeth, Benjamin S., *The Transformation of American Air Power*, Ithaca, New York: Cornell University Press, 2000.

Link, Maj Gen (Ret) Charles D., "Leading Airmen," *Aerospace Power Journal,* Vol. 15, No. 2, 2001, pp. 7–12.

Love, W. Dwight, "Preview of TCAS II Version 7," *Air Traffic Control Quarterly*, Vol. 6, No. 4, 1998, pp. 231–247. Online at http://www.mitre.org/work/best_papers/best_papers_98/loves/loves.pdf as of March 16, 2005.

Lyons Terence J., Norbert O. Kraft, G. Bruce Copley, Clark Davenport, Kevin Grayson, Heidi Binder, "Analysis of Mission and Aircraft Factors in G-Induced Loss of Consciousness in the USAF: 1982–2002," *Aviation, Space, and Environmental Medicine*, June 2004, Vol. 75, No. 6, pp. 479–482.

Martin, Dr. Elizabeth L., *Night Vision Goggle Training in SUPT and IFF: A Feasibility Study*, Mesa, Ariz.: Air Force Research Laboratory, March 2004.

Milani, Ian, telephone interview with John A. Ausink (Arlington, VA) and Michael Thirtle (Chicago, IL), August 17, 2004.

Mission Analysis Study Group, *Mission Analysis on Future Undergraduate Pilot Training: 1975 Through 1990, Volume I, Final Report*, AFSC-TR-72-001, Randolph AFB, Tex., January 1972.

Moorman, Gen Thomas S., Jr., Vice Chief of Staff, U.S. Air Force, "The Challenges of Space Beyond 2000," Remarks to the 75th Royal Australian Air Force Anniversary Airpower Conference, Canberra, Australia, June 14, 1996.

Ochmanek, David, *Military Operations Against Terrorist Groups Abroad: Implications for the United States Air Force*, Santa Monica, Calif.: RAND Corporation, MR-1738-AF, 2003.

Pae, Peter, "Unmanned Aircraft Gaining the Pentagon's Confidence," *Los Angeles Times*, August 7, 2004, p. C-1.

Palumbo, Col Frank, "Flying Training 2020," briefing, Randolph AFB, Tex., October 17, 2003.

Pyles, Raymond A., *Aging Aircraft: USAF Workload and Material Consumption Life Cycle Patterns*, Santa Monica, Calif.: RAND Corporation, MR-1641-AF, 2003.

Roche, James G., Secretary of the Air Force, "Lessons Learned from OEF," *Air Force Policy Letter Digest*, December 2002.

———, "The Future is Now—A Perspective on Air and Space Power in 21st Century Conflict," Remarks at the Command Chief Master Sergeant Conference, Gunter Annex, Maxwell Air Force Base, Ala., April 25, 2003a.

———, "Integrating Space into Joint Warfighting: Continuing the March," Speech to the National Reconnaissance Office Space Warfighter Conference Dinner, July 14, 2003. Online at http://space.au.af.mil/documents/sph2003_25.htm as of March 15, 2005.

Sanger, David E., and Eric Schmitt, "C.I.A. Tip Led to Strike on Baghdad Neighborhood," *New York Times*, April 8, 2003, p. 1.

Secretary of the Air Force, "Report of the Secretary of the Air Force," *Annual Report to the President and Congress,* 2002.

Shircliffe, David W., "Concepts and Programs for Future Undergraduate Pilot Training," AETC History Office, September 1975.

SPG Media PLC, "T-50 Golden Eagle Jet Trainer and Light Attack Aircraft, South Korea," *Airforce Technology*. Accessed September 23, 2004. Online at http://www.airforce-technology.com/projects/t-50 as of March 16, 2005.

Taylor, William W., J.H. Bigelow, S. Craig Moore, Leslie Wickman, Brent Thomas, and Richard Marken, *Absorbing Air Force Fighter Pilots: Parameters, Problems, and Policy Options*, Santa Monica, Calif.: RAND Corporation, MR-1550-AF, 2002. Online at http://www.rand.org/publications/MR/MR1550 as of March 16, 2005.

Taylor, William W., S. Craig Moore, and C. Robert Roll, Jr., *The Air Force Pilot Shortage: A Crisis for Operational Units?* Santa Monica, Calif.: RAND Corporation, MR-1204-AF, 2000. Online at http://www.rand.org/publications/MR/MR1204/ as of March 16, 2005.

Uhlarik, John and Doreen A. Comerford, *A Review of Situation Awareness Literature Relevant to Pilot Surveillance Functions,* Federal Aviation Administration, Office of Aerospace Medicine, DOT/FAA/AM-02/3, Washington, D.C., March 2002. Online at http://www.hf.faa.gov/docs/508/docs/cami/0203.pdf as of September 16, 2004.

U.S. Air Force, "U.S. Air Force Cost and Planning Factors," Washington, D.C., Air Force Instruction 65-503, 1994.

———, "RQ-/MQ-1 Predator Unmanned Aerial Vehicle," *Air Force Link*, July 2001. Online at http://www.af.mil/factsheets/factsheet.asp?fsid=122 as of September 24, 2004.

———, *Report of the Aging Aircraft Technologies Team on USAF Aging Aircraft Structures and Mechanical Subsystems*, 2002. For official use only; distribution limited to U.S. government agencies and their contractors.

———, "Global Hawk," *Air Force Link*, April 2003a. Online at http://www.af.mil/factsheets/factsheet.asp?fsid=175 as of September 24, 2004.

———, "Joint Specialized Undergraduate Pilot Training," *Air Force Link*, April 2003b. Online at http://www.af.mil.factsheets/factsheet.asp?fsID=181 as of March 16, 2005.

———, "T-38 Talon," *Air Force Link*, 2003c. Online at http://www.af.mil/factsheets/factsheet.asp?fsID=126 as of March 23, 2005.

———, *USAF Almanac 2003*, 2003d. Online at http://www.afa.org/magazine/May2003/default.asp#almanac as of March 16, 2005.

———, "80th Flying Training Wing, Euro-NATO Joint Jet Pilot Training Fact Sheet," 82nd Training Wing Public Affairs Office, 2004a.

———, "AWACS Mission Training Center," *Distributed Mission Operations: AWACS Program*, 2004b. Online at http://dmt.wpafb.af.mil/Documents/awacsdmt.htm as of September 23, 2004.

———, *Posture Statement 2004*, 2004c. Online at http://www.af.mil/library/posture/posture2004.pdf as of March 15, 2005.

———, "The U.S. Air Force Transformation Flight Plan, 2004," Washington, D.C., USAF Future Concepts and Transformation Division (HQ USAF/XPXC), 2004d.

U.S. Air Force Scientific Advisory Board, *New World Vistas: Air and Space Power for the 21st Century*, Washington, D.C., 1995.

U.S. Government Accountability Office, *Trainer Aircraft: Plans to Replace the Existing Fleet*, GAO/NSIAD-89-94, 1989.

———, *Tactical Aircraft: Modernization Plans Will Not Reduce Average Age of Aircraft*, GAO-01-163, February 2001.